T0331268

The
Thirteenth
Labor

WORLD FUTURES GENERAL EVOLUTION STUDIES
A series edited by Ervin Laszlo
The General Evolution Research Group
The Club of Budapest

See the back of this book for other titles in World Futures General Evolution Studies.

The Thirteenth Labor

Improving Science Education

Edited by

Eric J. Chaisson

Tufts University
Medford, Massachusetts, USA

and

Tae-Chang Kim

Institute for the Integrated Study
of Future Generations
Kyoto, Japan

Gordon and Breach Publishers

Australia Canada China France Germany India
Japan Luxembourg Malaysia The Netherlands
Russia Singapore Switzerland

Amsteldijk 166
1st Floor
1079 LH Amsterdam
The Netherlands

British Library Cataloguing in Publication Data

The thirteenth labor : improving science education – (World
    futures general evolution studies ; v. 15 – ISSN 1043–9331)
    1. Science – Social aspects – Congresses 2. Science – Study
    and teaching – Congresses 3. Communication in science –
    Congresses 4. Science – Public opinion – Congresses
    I. Chaisson, Eric II. Kim, Tae-Chang
    500

    ISBN   90-5700-538-7

    ISBN-13: 978-9-0570053-81

Contents

Introduction to the Series

The *World Futures General Evolution Studies* series is associated with the journal *World Futures: The Journal of General Evolution*. It provides a venue for monographs and multiauthored book-length works that fall within the scope of the journal. The common focus is the emerging field of general evolutionary theory. Such works, either empirical or practical, deal with the evolutionary perspective innate in the change from the contemporary world to its foreseeable future.

The examination of contemporary world issues benefits from the systematic exploration of the evolutionary perspective. This happens especially when empirical and practical approaches are combined in the effort.

The *World Futures General Evolution Studies* series and journal are the only internationally published forums dedicated to the general evolution paradigms. The series is also the first to publish book-length treatments in this area.

The editor hopes that the readership will expand across disciplines where scholars from new fields will contribute books that propose general evolution theory in novel contexts.

Preface

For two days in the fall of 1995, a highly diverse and international group of invited guests came together at the American Academy of Arts and Sciences in Cambridge, Massachusetts to discuss a broad and timely topic of interest to all people. The first Boston Forum on "Science, Education, and Future Generations" was designed to explore how we might improve science literacy among future citizens—and generally how we can foster a more humane, globally oriented society, given, or perhaps despite, the technological democracy in which most of us live. The objective was to convene a decidedly interdisciplinary cluster of broad thinkers to address the issue of how we can help build a better future for all humankind—and specifically what role science and technical education have, if any, in helping to ensure that better future.

Our spirited debate and deliberations were guided by an intentionally relaxed agenda, finding only fleeting foci within our wide-ranging conversations. This minimal structure was deliberate among the organizers; for we aimed to achieve a liberal interchange among a small gathering of eclectic individuals, none considered "expert" at the interface of science education and future studies, who had thought deeply and written about the issues at hand for many years. The intent was to seek fresh approaches, interdisciplinary cross talk, a suite of new ideas, by re-creating a forum of old—a free and open interaction among a score of wise men

and women from very different backgrounds and disciplines—
scientists, educators, philosophers, diplomats, futurists and
theologians, among others, including Nobel laureates, businessmen
and high-school teachers.

No papers were read at the Boston Forum, no planned speeches
given. Our idea was to allow each participant the opportunity to
have his or her say with maximum ease and comfort, yet to enable
others to interrupt the flow of discourse if so moved. The forum
itself was facilitated by using the council room at the American
Academy, where the twenty-two active participants sat around a
large octagonal table without any obviously preferred or central
location. Behind us, in an "outer orbit," a dozen observers—mostly
master pre-college teachers—entered the conversations sparingly,
just enough to keep our discussions relevant and useful to real-
world objectives. All sessions were translated simultaneously into
Japanese and English and video taped by NHK, Japan's public-
broadcasting system.

The resulting intellectual atmosphere was stimulating and
productive, the exchanges fruitful, indeed infectious at times.
However, and intentionally, there were no conclusions drawn, no
recommendations rendered by the group as a whole. We wanted
to brainstorm without restrictions; think innovatively and creatively
about the issues at hand; draw from each other frank and candid
feelings that otherwise might not be expressed at a more formal,
academically oriented meeting with strong chairpersons, time limits
and prepared lectures.

The four sessions comprising the daily workshops were loosely
arranged, each around a topic that we, as forum organizers, felt
was central to the global problem of science illiteracy.

The first session, featuring introductory remarks by all, held
high the common query: What can science education do now for
the well-being of future generations?

The second session explored how we might achieve a better
balance between the broader, integrated, interdisciplinary view of

science that will likely better serve future generations and the current highly specialized model under which we all now labor.

The third session examined novel, reformed educational programs and activities needed to better inform non-scientists about our technological world, as well as more systemically train scientists and technologists of future generations.

And the final session sought to identify action items and specific plans to aid future generations in creating a more technically literate yet humanely oriented society—the kinds of things we can do now to lay the foundation for science and science education in the new world order.

Many of the papers collected here were written in light of the forum's lively discussions, often incorporating aspects of the wide spectrum of ideas and arguments expressed during the two days of nearly non-stop interchange on the ways and means that science and science education can better serve future generations. Daily working sessions continued during long breaks, while walking through Norton's Woods at the academy, and on into the evening at the nearby Harvard Faculty Club, aided and abetted by two provocative after-dinner speakers: Hungarian-Italian systems philosopher Ervin Laszlo and Swiss Nobel biologist Werner Arber.

The main title of this volume, *The Thirteenth Labor*, we took from an especially insightful essay by the American Nobel chemist Dudley Herschbach. Speculating about how the mythological Hercules might have tackled a hypothetical, monumental task, or "thirteenth labor," such as weighing the Earth's atmosphere, Herschbach argues that what is needed is a novel approach, a whole new set of ideas. And that is the case, we feel, regarding the task at hand; to improve science literacy and the public understanding of science *tomorrow*, innovation, creativity and even maverick ways of exploring the problem are needed *today*.

The Boston Forum was only one of several working meetings occurring throughout the world in 1995, under the sponsorship of the Future Generations Alliance Foundation, a relatively new philanthropic organization dedicated to making planet Earth a more secure and hospitable home for present and future generations. Founder and chairman of this foundation is a remarkable Kyoto, Japan businessman, Katsuhiko Yazaki, whose aim is to affect change by means of innovative educational and social programs. Each forum addressed a different topic, such as environmental pollution, agricultural reclamation and public policy issues, their common denominator being global import; it was our task in this meeting to consider issues concerning science, technology and technical education.

In addition, the Boston Forum was pleased to have the support of the Fondation H. Dudley Wright, a Geneva-based institution dedicated, in part, to the creation and sharing of novel instructional techniques and interdisciplinary resources for pre-college science teachers. This foundation, one of whose principal vehicles of dissemination is the Wright Center for Innovative Science Education at Tufts University, underwrote an opening reception at the Hosmer House in Concord, Massachusetts. Administrative and organizational aid for the forum were provided and ably managed by Wright Center program coordinator Ellen Boettinger-Lang.

We offer this volume of essays in the spirit of bettering East–West cooperation on global issues of importance to all humankind.

Eric J. Chaisson
Tae-Chang Kim

Boston Forum Attendees

PARTICIPANTS

WERNER ARBER   biologist, Biozentrum, University of
              Basel, Switzerland*

WILLEM BROUWER   retired optician, Lexington,
                 Massachusetts, USA*

ERIC J. CHAISSON   astrophysicist and director, Wright
                   Center, Tufts University, Medford,
                   Massachusetts, USA*

DAVID CHEN   biophysicist and science educator, Tel Aviv
             University, Israel*

VILMOS CSÁNYI   geneticist, Oetvos University, Hungary*

E. JULIUS DASCH   geologist and head of NASA's Space
                  Grant Program, Washington, DC, USA*

FREEMAN DYSON   physicist, Institute for Advanced Study,
                Princeton University, New Jersey, USA*

RIANE EISLER   cultural historian, Center for Partnership
               Studies, California, USA*

DAVID ELLIS   chemist and president, Boston Museum of
              Science, Massachusetts, USA*

ANDREW FRAKNOI   astronomer and educator, Foothill
College, California, USA*

URSULA W. GOODENOUGH   biologist, Washington
University, St. Louis,
Missouri, USA*

DUDLEY HERSCHBACH   chemist, Harvard University,
Cambridge, Massachusetts, USA*

KIDOU INOUE   Zen master and theologian, Kyoto, Japan

TAE-CHANG KIM   head of Institute of Integrated Study of
Future Generations, Kyoto, Japan

ERVIN LASZLO   philosopher and systems theorist, Pisa, Italy*

DAVID LOYE   psychologist, Institute for Futures Forecasting,
California, USA*

LOYAL RUE   philosopher and theologian, Luther College,
Iowa, USA*

BRIAN SWIMME   mathematician, California Institute of
Integral Studies, San Francisco, USA*

RONALD K. THORNTON   particle physicist and science
educator, Tufts University,
Medford, Massachusetts, USA*

JANET WARD   cognitian, and head of Scientist as Humanist
Project, New Hampshire, USA*

URI WILENSKY   mathematics/media educator, Tufts
University, Medford, Massachusetts, USA

KATSUHIKO YAZAKI   chairman, Future Generations
Alliance Foundation, Kyoto, Japan

---

* Also essayists in this volume.

OBSERVERS

SCOTT BATTAION   biology teacher and Wright Fellow, California, USA

MARY ANNE CHURCH   biology teacher and Wright Fellow, Washington, USA

NEIL GLICKSTEIN   marine biology teacher and former Wright Fellow, Massachusetts, USA

ILEANA JONES   physics/astronomy teacher and former Wright Fellow, Massachusetts, USA

JAMIE LARSEN   environmental teacher and Wright Fellow, Arizona, USA

GEORGE LEONBERGER   chemistry/astronomy teacher and Wright Fellow, Texas, USA

JAMES MacNEIL   geology and high-technology teacher, Massachusetts, USA

STEVE METZ   interdisciplinary science teacher, Massachusetts, USA

JANET KRESL MOFFAT   biology teacher and former Wright Fellow, Massachusetts, USA

CHRISTOPHER RANDALL   earth science teacher and former Wright Fellow, Massachusetts, USA

WALTER STROUP   physics teacher and former Wright Fellow, Massachusetts, USA

RONNEE YASHON   educational coordinator, Wright Center, Tufts University, Medford, Massachusetts, USA

The
Thirteenth
Labor

1

Toward a Scienceless Society?

Eric J. Chaisson

Several months before convening the Boston Forum, I wrote a "working paper" that was intentionally controversial. It set forth a somewhat extreme view of how I regard science and technology potentially affecting the lives and attitudes of future generations. I sent it to the Forum participants, hoping to get their intellectual juices flowing well before the meeting, indeed aiming to provoke them into thinking broadly about the central issue at hand: What can science education do now for the well-being of future generations?

My intent, openly expressed then and now, is to challenge my colleagues to reform the ways that both the educational community trains our teachers and the scientific community interacts with the general public—and especially to encourage all of us to consider science and science education in a more integrated, interdisciplinary manner. These, I believe, are among the keys to solving the growing problem of science illiteracy in our technological world today.

One note on a definition, as pertains throughout this collection of essays. By "science education," we mean both science and mathematics instruction, including insights from engineering as well as from the humanities and social studies, for all these are areas where much improvement is needed in the quality of teaching, from kindergarten through graduate school. We also take science education to encompass outreach activity meant to disseminate good, solid science to the public in an accurate and understandable manner.

A Subjective Prognosis

Although I do not claim to have 20/20 vision even as far out in front of us as 2020 A.D., I do suggest that, in the near-term, we are in danger of creating a "scienceless society." By using this revolting phrase, I mean (at least figuratively, and possibly literally) that one of these days, perhaps soon, the lay public will begin knocking on the front doors of science departments, informing those of us within that our citizens neither understand what we are doing nor intend to support our work any longer. The result will likely be a steady, perhaps nearly irreversible, trend toward a society mostly lacking in scientific curiosity, the likes of which the world has not experienced since the thousand-year doldrums of the Middle Ages. That trend may have already begun in the 1990s.

In saying this, I am not claiming that we are about to abandon technology and crawl back into the bush or the cave; by contrast, all things technological, applied, and economically viable will be roundly embraced. Yet science itself—knowledge for the sake of knowing, the drive to understand pure and simple, the study of Nature for the sheer beauty of it all—will likely become increasingly less valued in these days of federal deficits and rising fundamentalism. Already Americans, especially, are opting out of science in alarming numbers, the impression being that science is just not worth bothering about.

The issue I raise is hardly a new one. But I am not here referring casually to the growth of the celebrated culture gap; not merely concerned that our society might be unfortunately fragmenting, as C.P. Snow warned us decades ago, into dual camps of the intellectual elite and those who aren't. Rather, I sense a dramatic widening of that gap, especially as pertains to technical matters, to the possible detriment of all humanity.

Most of us live in a technological democracy, yet the average citizen is losing interest in (and perhaps developing a fear of) science and technology. To my mind, that's because the public doesn't understand (or even appreciate) science and technology—and especially doesn't understand (or trust) what we are doing with these twin endeavors. I am not concerned about subtleties in the understanding of everyday phenomena, like the causes of the seasons on planet Earth, for example, or the way that speech passes through air. Rather, I refer here to more basic misunderstandings, such as these sobering statistics uncovered recently during a U.S. government survey:

- nearly one-third of the adult population is unaware that Earth goes around the Sun.
- another third of all adults does not know that Earth takes one year to orbit the Sun.
- roughly half of elementary science teachers think that sound travels faster than light.
- a majority of adults believe that humans coexisted with the dinosaurs.
- many inner-city science teachers are unsure if atoms are real, or if stars are hot.

The upshot is an undermining of our economy, which today depends on advanced technology and industrial motivation. A further result is an erosion of our technological workforce, which might become a threat to our democratic way of life. After all, and forevermore, we shall, as voters, be confronted by global issues of a technical nature (what with environmental pollution, climatic change, nuclear waste, genetic engineering, health care, etc.), and unless we can sort it all out before entering the voting booth then we might well be casting away our future. The question before us is this: Why should a civilization that doesn't understand some of the basics of science, or at least feel comfortable with the advance of technology, want to embrace our headlong thrust toward an increasingly technical and complex world, let alone build scientific and technological gadgets to continue the exploration of space or the hunt for another quark?

The Source of Science Illiteracy

Many statistics can be cited, like those few noted above, to imply a poor understanding of science among the world's populace today. And it is not a problem confined to the United States, as one of the Japanese speakers at the Boston Forum made clear regarding the poor state of science literacy in his country as well. Suffice to note that countless citizens, even in the industrialized nations, cannot balance a check book or even add correctly a column of numbers, let alone program a personal computer or operate a video cassette recorder. A significant fraction of the current generation of global adults—again, I stress, many of them voters within technological democracies—are effectively illiterate as pertains to rudimentary socio-technical issues involving energy resources, environmental protection, cyberspace usage, and the like.

To my mind, the source of the problem is straightforward: today's typical pre-college teacher, with some glowing exceptions, does not understand science well enough to teach it. Most classroom teachers are skilled in many things, such as the theory of learning, methods of teaching, critical thinking, group dynamics, classroom discipline, and psychological analysis; and most of these assets are doubtlessly needed for a vibrant educational experience. But that same classroom teacher is largely lacking in the fundamentals of science, not to mention having any kind of feeling for what it's like to *be* a scientist or to *do* science. In simplified terms, for this is not meant to be an indictment of what has gone wrong—rather a brief essay to suggest ways to help correct it—there are two reasons for the scienceless science teacher, hence for the current state of technophobia in our society today.

First, the professional educational community has, during the last several decades, produced two generations of pre-college teachers who know very little science. I do believe that it is as simple as that; to be sure, most experienced teachers agree. Until recently, Education Schools associated with universities graduated, and states of the U.S. certified, students who then became science teachers skilled, as noted above, in many areas other than science. Now, owing to the looming crisis in science education, governments—mostly at the state and local level—are demanding that new teachers of science receive significant training in science. Consequently, to gain appropriate certification to teach (at least in public schools), many of today's pre-service teachers now study in college the subject matter that they actually intend to teach later in life.

This common-sense notion would surely seem a positive step forward, yet the reader might be surprised how much the old guard educators (who know little science) have been reluctant to embrace it. Educational inertia is no more prevalent anywhere in society than in the Education Schools themselves. Mostly because they have been blamed by much of society for the ills of society, and partly because they deeply resent any advice from the outside, Education Schools have become decidedly defensive, insular, and largely unreceptive to proposals for change, especially from the scientific community.

A second broad cause of today's science illiteracy is the scientific community itself, which has, during roughly the same, two-generational period, abrogated its responsibility to disseminate science, especially among our children's educators. Since the end of the Second World War and the rise of government-sponsored grantsmanship, scientists have unquestionably emphasized research over teaching, discovering over sharing. Excellence in the lab is clearly rewarded more by universities than are efforts to teach science well. Nearly a half-century of federally funded research has led to some magnificent advances in the world of science and technology, many of which make our everyday lives more comfortable and productive, but the decoupling of the scientific community—in government, industry, and academia—from the pre-college domain has caused an alienation from science among elementary and secondary school teachers that has worsened throughout the past few decades.

Nor have scientists done well in the wider, public domain beyond the formal educational system. We have simply not conveyed the value of our work to the voters who choose the politicians. Only now that technical illiteracy has escalated worldwide to potentially threaten the self-preservation of the scientific community are colleague scientists showing sudden interest in helping to ameliorate the problem. Yet here arises another, potentially more disturbing, issue and I refer not merely to the lukewarm expressions of concern among many leading science societies and organizations.

We are now beginning to see the early strains of the politicization of science. As research scientists joust for decreasing funds in budget-balancing and economically lean times, many of them, and especially their parent agencies and institutions, are now resorting to hyperbole and overselling to keep the flow of funds and favor coming. The inevitable result is both an exaggeration of scientific findings and a worsening of the public's understanding of what actually has been accomplished scientifically, ironically often at taxpayer expense.

The National Aeronautics and Space Administration (NASA), a government agency with which I have had close dealings for many years, is especially good at this bad habit. As NASA's budget shrinks all too fast, mostly owing to its own technical shortcomings and managerial confusion, public-relations announcements stream forth, falsely proclaiming ever-grander discoveries that almost weekly are said to change our view of the cosmos. Consider just the single (budget-battling) month of January 1996, during which I am writing this article, when NASA's science managers claimed "major discoveries" for a Universe that is vastly more populated with galaxies, a planet Jupiter that is hotter, windier and lacking in helium, a planet Saturn that has several new moons, and the first photo of the surface of another star beyond the Sun—claims which are mostly wrong and at best inaccurate.

Equally distressing is the growing opinion of some scientists that this kind of "selling of science" is needed, both to gain credit in the present and to win funding in the future. To my mind, it is bad enough for colleague scientists to tolerate politely exaggerated results of science findings, but unacceptably worse to witness increasing numbers of scientists lowly urging that we make such extravagant claims as a matter of course to preserve the base of research funding. And, of course, all this science hype reaches new heights of disinformation in the hands of an indiscriminate (and often unknowing) news media, which have a vested interest in the kind of sensationalism that sells all too well in a gullible society having hardly more than a clue whether to believe it or not.

Treatment for the Ailment

The Boston Forum identified and discussed a number of useful approaches toward bettering both science education and public attitudes regarding science and technology. Every attempt was made to specify tasks and action items that participants (and readers of this volume) can incorporate into their daily routines, all with the aim of improving science literacy now and in the future.

In their keynote addresses, Werner Arber of Basel and Ervin Laszlo of Pisa appealed to both the practical and philosophical, citing respectively proven programs that have measureably improved science education as well as the rights and responsibilities of scientists to future generations. Arber, in particular, made clear the need for more transdisciplinary work in all of education, and not just science education, stressing the value of experts having a deep appreciation of subject matter beyond their specialties. This Nobel-biologist describes in his paper a half-dozen specific programs in science education for non-scientists and in science communication for the general public that have worked well during the last decade at the University of Basel. These programs include a two-hour-long slot on the university-wide weekly calendar reserved exclusively for cross-disciplinary seminars, an off-site institute in the Swiss Alps dedicated solely to interdisciplinary course work for credit, and a Eurodoctorate in Biotechnology that incorporates disparate fields such as economics, bioethics, patent law, management, and environmental science, as well as courses in the natural sciences.

Somewhat in contrast to this practical approach, though still embracing interdisciplinarity as central, Laszlo took the "high road," arguing that science has become a major force shaping contemporary society, and it is for this reason alone that the science community needs to be more conscious of science's impact on humanity. Appealing to the systems philosopher within him, Laszlo raises in his paper the neglected, but not negligible, implications of the contemporary vision projected by science of humankind in the universe. "Whether that vision is true or false, constructive or threatening, it shapes our perceptions, colors our feelings, and impacts on our assessment of individual worth and social merit." Like the main benefactor of the Boston Forum, Katsuhiko Yazaki of Kyoto, Laszlo argues that the best route to our future survival requires a change in our awareness of the world around us to include value, heart, warmth and soul—qualities capable of painting a more friendly and humanizing portrait of who we are and whence we've come. Rather than conveying a dehumanized vision of the world, dry and abstract, reduced to numbers and formulas without feeling and value, it is up to the scientists themselves to articulate a more positive picture of our findings, thus describing for all the public that "stupendous view of ongoing creativity" seen throughout Nature.

Interdisciplinarity in science and science education, not long ago the scorn of highly specalized academia, resounded throughout many of the arguments heard at the Boston Forum. Futurist Riane Eisler urges us to adopt a broader approach to science education, "one that teaches us from early childhood on to think of science . . . in its social and ideological context"; appealing to a partnership of men and women, she urges us to create a new form of science education having a more feminine ethos of caring, values, and compassion, thereby utilizing science and technology in more humanistic ways. Likewise, Brian Swimme, a mathematician by training, seeks that larger, warmer, more noble science story, stating that, not merely a collection of facts, science should be a student's guide to a grand world-view, including, if possible, meaning, purpose and value; he sees the cosmological perspective as one to which all modern scientists can objectively subscribe, yet the meaning and purpose of it being a subjective outgrowth of an individual's reflection upon that cosmology—to Swimme, a vital educational component of the next millennium.

Freeman Dyson, in a wonderful essay about three unrelated and contrasting characters of American, French and Russian history, offers that it is the degree of social justice that most dictates whether we have good or bad science education; the renowned physicist from the Institute for Advanced Study maintains that our schools need both Napoleonic discipline and Tolstoyan freedom, the former to train the technical experts who can run a technological economy and the latter to ensure that all citizens are culturally and scientifically literate enough to put the benefits of technology to the kind of good use that, ironically, labor leader Samuel Gompers foresaw about a century ago.

In a delightful paper that captures the wisdom of having taught chemistry for forty years, Nobel Laureate Dudley Herschbach implores us to embrace a "liberal science" approach to science education, akin to the broad perspective nurtured in the liberal arts—to cherish "the human advanture of intellectual exploration . . . ultimately achieving wondrous insights." As with the best papers in this collection, he offers specific remedies—a 10-point agenda of proven ideas—for better assimilating science and mathematics into our general culture.

Astronomy educator Andrew Fraknoi also spoke mainly to the issue of societal woes, noting that a by-product of poor science education (and a sensationalizing media) is an adult population unable to think skeptically—the result being a rising level of gullibility among citizens unsure what to make of new-age cults, fundamentalist religions, profit-making charlatans, among a long list of other pseudo-science groups, all of whom tend to make today's socio-technical issues seem ever more complex. Biologist Ursula Goodenough also pitched her remarks toward society, contending as unrealistic the idea that the average citizen will be able to make valid technical decisions in an increasingly complex world of the twenty-first century; rather, she says we must rely on experts, who, however, increasingly have their own subjective agendas, including deception.

Among other common threads woven throughout some of the essays in this collection, the reader will encounter the voguish topics of complexity and chaos, here pertaining to modern society, and once again pleas for intellectual cross-breeding if we are to sort it all out. The biophysicist David Chen of Tel Aviv well outlines the digitization of the knowledge age—a whole new era of human civilization now bursting upon us; he helps us understand how yet another gap, between the info-rich and the info-poor, continues to widen the world over, yet even the info-rich of the West struggle with the cognitive demands to decipher the myriad complex phenomena around us. The geneticist Vili Csanyi of Budapest spoke of an approaching bifurcation within the highly unstable state of humanity, echoing the social psychologist David Loye in calling for a greater sense of moral consciousness to help guide us onto a path of dynamic self-improvement—a kind of personalized education. Throughout the Forum, Csanyi and Loye repeatedly championed the moral imperative needed to reconceptualize science education.

Janet Ward, a humanist from small-town New Hampshire, raised again the big issue of connectedness—an "authentic and growing appreciation that everything is a part of a cosmic nexus"; her message, however and somewhat unorthodox, is that a vital part of this newly recognized interrelatedness is the inherent notion that the path taken by humankind at this particular turning point is likely more a matter of subjective wisdom than of objective knowledge. Is the will more important than the facts? Optician Willem Brouwer also expressed a view toward alternative schooling, arguing that formal and rigid learning might be a mistake; rather, schools should allow pupils much more freedom and choice, granting children the means for learning in a way that is unstructured yet closely connected with their real-world curiosity of the moment.

Nearly all the attendees of the Boston Forum offered specific recipes for improving science education, and three of them reviewed activities known to work. Chemist and museum director David Ellis presented a case that science centers (museums, aquariums, zoos, etc.) already contribute significantly to the struggle for science literacy and are likely to become increasingly important players in the adjunct area of informal science education, especially perhaps in conjunction with the Internet and its World Wide Web. NASA administrator Julius Dasch, a geologist by training, summarized his positive experiences with "gateway" courses designed to reach maximum numbers of university students in a relatively painless way—the objective being to advance, aggressively yet enjoyably, scientific and technical literacy among large segments of students who otherwise might not study much science at all. And Ronald Thornton, particle physicist turned educator, described his group's success in using non-traditional methods to teach principles of physics; the use of simple pedagogical tools (personal computers, motion sensors, etc.) and hands-on activities (that mimic real-life experiences) have dramatically improved student understanding of the basic concepts.

Not by any means least, religions of both the East and West were represented at the Boston Forum. In the concluding paper that juxtaposes this opening piece, Christian theologian-philosopher Loyal Rue of America's midwest took direct aim on my strawman prospect of a "scienceless society," proposing, instead of more tactical talk of curriculum reform in the schools, that we strategically generate a new philosophical myth—an integrated world-view that incorporates within it the most powerful kind of science education for future generations: a narrative scientific framework understandable to all and a mental attitude conducive to life-long learning, a consequence of which would likely be a "science-friendly society."

Conclusion

What is most needed right now is a greater sense of balance, a reordering of our educational priorities—in two ways. First, the professional Education Schools (which are effectively the sole trainer of our childrens' teachers) need to recognize that the next generation of teachers must know and appreciate better the sum and substance of science—and not merely the learning theories, behavioral attitudes, or historical objectives, which are among the central topics those schools now so heavily emphasize. Short of such genuine reform, whole other means—beyond the Education Schools—must be devised to train and retrain our science teachers, for without more knowledgeable teachers in the subject matter of science itself, the lingering problem of technical illiteracy cannot be solved regardless of what else is done.

And as a second balance, the scientific community needs to recognize that sharing is as worthy as discovering, teaching as honorable as researching, synthesizing as valuable as specialized work. In today's world of an active media and public accountability, the capacity to explain science well, accurately and without hyperbole, and the garnering of public trust that results therefrom, are vital parts of the scientific enterprise. Otherwise, as scientists, alongside today's ill-trained educators, we shall continue to be the greatest cause of a technically illiterate population.

Perhaps it would be wise, just now, to pause for a few years, a decade, a generation—maybe until 2020 A.D.—and truly revitalize our science and technology educational base, from grade school to grad school. Perhaps it would be reasonable to readjust the educational fulcrum for the research and teaching of science, as well as for the public funding and understanding of science. Then, after we have raised the technical literacy of all our citizens, we can confidently move forward toward making real and relevant some of the wondrous things that science and technology might provide for the betterment of all humankind.

2

Transdisciplinarity in Science Education and in Science Communication

Werner Arber

In a world in which human society is steadily confronted with complex problems, interdisciplinary collaboration between people with different professional backgrounds is needed more and more. Institutions of education, and particularly of higher education, should thus periodically evaluate their study plans with the goal to offer to the next generation the most appropriate education to prepare them for both their professional lives and their lives as citizens. Such evaluation should take into account the following three points:

First, education should offer the opportunity to acquire profound knowledge in the discipline chosen by the person. But in addition to the anchorage in this discipline, the student should develop interests for other fields of study and get a chance to experience interdisciplinary work in a context requiring, besides his or her own field of expertise, a number of other disciplines because of the complexity of the questions to be solved. Second, in view of the high speed of cultural as well as technological progress and taking into account emerging new needs, particular attention should be paid to the principle of recurrent education, of life-long learning. Already during his or her basic training, the student should understand that education should not be considered complete with the final examination, but that recurrent education will be needed to keep him or her up-to-date throughout his or her professional life. Third, I also consider a solid moral basis as a necessity for a person to carry out his or her professional duties with responsibility.

The listed requirements apply in principle to any professional education, although in differing degrees. They should have their impact at various levels of education from the primary school up to the university. Here, I will concentrate on education in natural sciences and I will especially refer to higher education at university level.

Science education is an important basis for our life in a technological world. It has its role not only for its contributions to technological progress, but also for the updating of our world view and culturally based beliefs. It is clear that a majority of persons will not become scientists themselves, but they should have the opportunity to understand the principles of science in order to follow and critically evaluate their progress. For example, a lawyer should have the means to grasp the essentials of scientific progress and to conduct interdisciplinary work in order to accomplish his professional duties. Similar arguments apply to a history teacher as well as to most other professionals.

It might be useful to give here a short definition of the terminology used. Transdisciplinarity means that a person goes beyond his or her own field of competence and deals with aspects such as the main contents, strategies or methods applied by other disciplines. In interdisciplinary work, a number of persons with different backgrounds, each solidly anchored in his or her own discipline, collaborate with each other in order to solve complex questions of common interest. Often they may share strategies originating from some of the participating disciplines. The terms multi- and pluri-disciplinarity are sometimes used with quite similar meaning as interdisciplinarity.

At the level of higher education in the sciences, it is the task of the universities to offer to their students a solid grounding in each of the ever-increasing number of disciplines engaged in the scientific endeavour. The quality of this education will be revealed by the later success of the graduates in their professional lives, especially when they are exposed to high competition at the international level. In addition to this disciplinary education, the students should have opportunities to keep their horizon wide by following transdisciplinary courses, which introduce essential questions dealt with and strategies used by other disciplines including those of the humanities and the social sciences. This transdisciplinary education is essential as a preparation for successful interdisciplinary collaboration, in which different disciplines with different languages and different strategies meet, whereby a mutual understanding often depends on the ease of the communication between the parties involved.

Such statements apply of course to any of the fields of higher education. It is important that general education in scientific fields not be limited to basic understanding of scientific facts at the level of the public school. Rather, it should be continued up to the level of higher education. This requires that excellent and accessible courses in scientific fields should be offered to all those university students not enrolled in studies of the discipline in question. These courses should consist of well selected examples to explain the scientific approach and they should include introductions into main strategies and methodologies applied as well as an explanation of specific terminologies employed. Debates on the general relevance of the topics dealt with are also important, since such a dialogue can stimulate the mutual interest of the participants. The main goal of this transdisciplinary education is to pave the road to true interdisciplinary work.

Science communication directed toward the general public is also a transdisciplinary educational activity. This important task is mainly accomplished by science journalists and other science communicators, but it should also be part of the tasks of scientists engaged in basic and applied research. The quality of science communication can widely influence the opinion of the general public

on sciences and their role in the cultural progress of human civili-
sation. General attitudes in this regard can have important feed-
back on the future prospects of scientific developments. The task
of transdisciplinary science education deserves thus careful attention and should not be left to chance events.

Let me now illustrate what I have so far said by reporting on
trans- and interdisciplinary activities in which I was myself more
or less strongly involved.

The development of the powerful tools of molecular genetics
would not have been possible without interdisciplinary collaboration. It started in the 1940s when scientists realized that bacteria
and even viruses have genes which determine microbial life processes, just as it had already been known for higher organisms. Soon
afterwards it was shown, using both chemistry and microbiology,
that deoxyribo-nucleic acid (DNA) is the carrier of genetic information. A few years later the double helix structure of DNA was
discovered by biophysical approaches taking into account relevant
chemical and biochemical data. This, in turn, opened the door to
decipher the genetic code. Microbial genetics of bacterial viruses
and of their host bacteria revealed the existence of restriction enzymes providing bacteria with a kind of immune system acting
against invading foreign DNA. These explorations led us to understand how some restriction enzymes cleave long filamentous
molecules of DNA into shorter, specific fragments. With the involvement of chemical methodology, such fragments were purified
and then used to analyse the specific sequences of the nucleotides
of which DNA is composed. On the basis of known sequences, it
became possible with the help of organic chemistry and enzymo-
logy to alter (mutate) a given sequence at some selected sites in the
DNA molecule, all with the purpose of investigating the biological functions of the segment of DNA in question. This method of
site-directed mutagenesis rapidly became an important strategy of
investigations by so-called reverse genetics. In classical genetics,
the scientists start to search for individual organisms with mutated

functions, which are manifested by their different phenotypes. The geneticist then studies transmission of the mutated traits and he also locates the site of mutation on the chromosome. In reverse genetics the process is indeed reverse, starting with a segment of DNA of unknown function. After site-directed mutagenesis and introduction of the mutant sequence into an organism one can look for the resulting phenotype and thus determine the segment's (or gene's) biological functions.

Practical applications of such acquired knowledge can now be considered. Thanks to these new approaches of molecular genetics, biotechnology has received new impulses and will certainly continue to do so in parallel to the increasing knowledge on the molecular basis of life's processes. These applications are of immediate relevance for many different disciplines, such as medicine, health care, agriculture, food technology and ecology. They also have their impact on the economy, ethics, social sciences, law and philosophy. Many fields of the humanities and social sciences are thus engaged, and it is clear that the specialists of these disciplines should be able to deal with questions on the basis of acquired knowledge of essential aspects of the biological sciences, in particular of molecular and cellular biology as well as genetics. Similarly, in order to be prepared for interdisciplinary dialogue, the biologists should be able to understand the reasoning and arguments raised by members of the humanities and social sciences. Hence, transdisciplinary education is essential for all partners involved in dealing with questions related to the assessment of gene technology and of the application of new findings emerging therefrom.

In order to prepare students properly for interdisciplinary work, various approaches have been followed with quite a good rate of success at my university in Basel, Switzerland. Let me report on the experience gained in the last 8 years.

In 1987 we installed a transdisciplinary teaching program aimed at students of all faculties. In this program, gifted university teachers were asked to present essential aspects of their own discipline to students of other disciplines. In general, one hour per week was devoted to a specific topic during one semester, and about 10 different topics were offered in parallel sessions, for which the university had reserved two hours from 4 to 6 p.m. on Tuesdays. This has the advantage that both the engaged teachers as well as the students having no other duties often found the time to attend one or two of the offered lectures.

In the same time slot, a number of interdisciplinary topics were addressed by a series of presentations given by speakers of different disciplines, often followed by a discussion. The lecture series of both types listed above were attended by about 12 percent of the enrolled students and all faculties were represented, although unevenly, with theology students reaching the highest attendance and medical students the lowest. The engaged teachers felt that the response by the students was generally quite good and stimulating.

It is well known that students are often forced to set priorities. Transdisciplinary courses are thereby often given low priority. But not so for transdisciplinary courses given as a block and ideally away from the home university. This has the advantage that participants attending a block course (usually lasting several days) remain on the spot and actively attend the lectures and discussion groups. Such block seminars have been organized since 1990 at a remote center near Sion in the Swiss alps — the Kurt Bosch Institute. Most of those who attend such interdisciplinary block instruction experience a very welcome enrichment of their education. The Kurt Bosch Institute addresses its teaching program to students of all Swiss universities and of a few universities of the neighbouring countries. Often a course can also accept older adults in search of trans- and interdisciplinary continuing education.

For the past few years, a special program of topics on environmental sciences has been offered to all students of the University of Basel and can be chosen by students at any level of education in parallel to their normal disciplinary curriculum. This program is called "Man, society and environment" and is divided into three sections. Section A is devoted to four blocks of introductory lectures about the environmental sciences and deals especially with questions related to environmental problems. Section B offers a relatively large number of transdisciplinary courses under the following five headings: Nature, perception, technology, development and ethics. Section C offers the students opportunity for intensive practical interdisciplinary work on several topics of choice. Students having received the required number of credit units can, after successfully passing an examination, obtain a certificate of complementary studies in environmental sciences. Alternatively, an increasing number of disciplinary curricula accept segments of these studies as part of the electives needed for graduation. In this way, an increasing number of students become familiar with at least some of the complex problems of today's society. The acquired knowledge will certainly help many of the enrolled students in their future professional activities.

It is true that many parts of curricula offered at universities are already of an interdisciplinary nature. I myself participate in the teaching program leading to a diploma in biotechnology, given in a common effort by 4 universities of the Upper Rhine valley and located in France, Switzerland and Germany. I am also involved in the planning of a widely coordinated Eurodoctorate in Biotechnology. For an appropriate education in biotechnolgy we consider not only various disciplines of the biological sciences, chemistry and physics as essential, but also a number of other fields such as economics, bioethics, patent law, management and environmental sciences, as well as the ability to express oneself in more than one major European language.

During my participation in a coordinated interdisciplinary research program on biodiversity for the last few years, I realized the immense educational value of such programs, especially for those students intending to carry out collaborative professional work. Various specialists of different disciplines of not only biological and earth sciences, but also of law, philosophy and economics, are commonly engaged in this program to deepen our knowledge on the origin, stability and loss of biodiversity. Within this wide context, my personal contribution centers on the study of molecular mechanisms of biological evolution, in particular the process of spontaneous mutagenesis and the dynamics by which natural selection acts on mixed populations of genetically different organisms. Among the general conclusions drawn from this research is the claim that, at least in microorganisms and probably also in higher organisms, specific sets of genes exist which can be considered as motors of biological evolution without, however, deterministically driving evolution towards a specific goal. By its relevance for our world view, this conclusion again is of a highly interdisciplinary nature and deserves attention among specialists of various disciplines.

With these selected examples, I have tried to support the view that successful solutions to the complex problems of modern society require interdisciplinary strategies. The opportunity to participate in interdisciplinary work should be provided by teaching programs at all levels, in particular within higher education. A prerequisite to obtain this ability is the acquisition of trans disciplinary knowledge in other essential disciplines. However, in highly competitive research, a solid anchoring in at least one discipline is essential. Thus, I offer a general conclusion: academia should aim to offer each student a basic education of high quality within his or her chosen discipline, and it should also offer the possibility of widening students' horizons by providing them access to other disciplines, as well as to interdisciplinary work undertaken by groups of experts dedicated to solving complex global problems.

3

The Neglected, But Not Negligible, Responsibility of Science to Society and to Future Generations

Ervin Laszlo

The responsibility of science and scientists to society and to future generations is indisputable. Science wields power. This power is not direct, for scientists are seldom in a position to make crucial decisions that affect the life and well being of people now and in the long term. It is real, nevertheless. Knowledge is power, wrote Francis Bacon already in 1597, almost a century before Newton. This still applies today, with the proviso that the knowledge in question either comes from the sciences or has some measure of scientific legitimacy. Science has become a major—perhaps *the* major—force shaping the contemporary world, and therewith the conditions under which we, and our children and grandchildren, will live our lives. This implies responsibility. Many aspects of this responsibility have been recognized, especially since the first atomic bomb exploded over Hiroshima, but some aspects of it have been relatively neglected. One such aspect, unduly and even dangerously neglected, is the topic of this study.

It has become increasingly clear that the evolution of contemporary societies is driven more by the social and technological spin-offs of scientific innovations than by the power and will of politicians. The breakthroughs of microelectronics have opened the information superhighway to global traffic and now bring to the fingertips of those who navigate it ideas and images on practically everything in every conceivable field of interest, from local gossip to global crises. The technical applications of information and control discoveries enable many people to reduce working hours and increase leisure-time—and require countless others to retrain or face marginalization and unemployment. Innovations in transport technologies enable massive streams of tourists and business people to travel anywhere on the six continents in a matter of hours in considerable comfort and safety. Breakthroughs in biotechnologies make possible the enlargement of the food supply and the extension of the human life span, with fresh cures to the many diseases that still inflict the human condition. Paradoxically, even the absence of all-out war is due in some measure to advances in science-based technology: modern weapons have become so powerful that they now endanger the potential victors themselves, and reduce the spoils of war to heaps of rubble that may be poisoned or radioactive to boot.

But atomic bombs, intercontinental travel, miracle drugs, and instant communication are not the only kind of consequences entailed by science for society. The technological aspects of science's impact are widely recognized, but there are also neglected, yet not negligible, aspects. These include the vision projected by science of man and the universe. Whether that vision is true or false, constructive or threatening, it shapes our perceptions, colors our feelings, and impacts on our assessment of individual worth and social merit. It enters into the set of ideas, emotions, values, and ambitions that we call human consciousness—the warp and woof of our immediate experience. The question is not whether science affects our consciousness; it is only whether it affects it for the better or for the worse—whether it helps us meet our goals and realize our dreams, or leads us down blind alleys and exposes us to shocks and surprises.

This aspect of science's impact on society is neglected only at our peril. As Katsuhiko Yazaki points out, there are three major obstacles as we try to deal with the problems that confront us today: the physical and technical limitations of our world, its social and economic structure, and the block in our consciousness to address these issues.[1] We must agree with Yazaki when he writes that the most formidable obstacle is in our own mind. Yazaki's insight, that the only way to our survival is a total change in our awareness, has been uppermost in my thinking for a number of years. It has been the central concern expressed in my recent book *The Choice: Evolution or Extinction*,[2] and was the principal motivation in my efforts to create the Club of Budapest, an association of world-renowned writers and artists dedicated to the furthering of planetary consciousness.[3]

The vision of themselves and of the world most people derive from science needs to be rectified. It is not the correct vision; and it is unduly negative. For society at large, science conveys a dehumanized picture of the world, dry and abstract, reduced to numbers and formulas without feeling and value, heart or soul. Even in the mind of the more science-minded segment of the population, the scientific vision is incomplete and in some respects inhuman. The universe is a soulless mechanism, and life in it is but a random accident. The essential features of living species, the same as of our own, result from a succession of accidental events in the history of biological evolution on Earth, while the unique features of the human individual are due to a fortuitous combination of genes to which this succession has led when he or she was born.

It is not surprising that many people conclude that the long series of accidents that produced each of us, together with the struggle for survival to which we are always exposed, have made us into bundles of egotism, indifferent to all that lies beyond the compass of our interests and separate from nature, and all that lies beyond the limits of our body and the ego that inhabits it.

The contemporary sciences, especially in the leading-edge fields of physics and biology, could and should tell a different story. In light of the latest developments, the universe is not a lifeless, soulless

aggregate of inert chunks of matter; if anything, it resembles a
living organism more than it does a dead rock. This "creative
universe" brings forth all of us, and all we see around us, in a
stupendous process of ongoing creativity.[4] There is no longer a
categorical divide between the physical world, the living world,
and the world of mind and consciousness. "Progressing along
the arrow of time," Chaisson writes, "we can now trace a thread
of understanding linking the evolution of primal energy into
elementary particles, the evolution of those particles into atoms,
in turn of those atoms into galaxies and stars, the evolution of
stars into heavy elements, the evolution of those elements into
the molecular building blocks of life, of those molecules into life
itself, of advanced life forms into intelligence, and of intelligent
life into the cultured and technological civilization that we now
share."[5]

Entire new fields of inquiry have been created to link disciplinary domains that were hitherto separated. There is, for example,
quantum biology, linking quantum physics and the life sciences,
and quantum brain theory and consciousness research, bridging
the gap between physics and the cognitive sciences. The theories
outlined by some investigators—David Bohm, Ilya Prigogine,
Henry Stapp, and myself among others—apply with equal cogency
to physical and biological, and to psychological and sociological,
phenomena. In the emerging paradigm, "matter," "life" and "mind"
are consistent elements within an overall process of staggering
complexity yet coherent order.

There are no good reasons why science could not evolve across
the disciplinary boundaries that have traditionally confined its
fields of investigation. These boundaries are not the works of
nature; they are the creations of scientific, above all, of academic,
communities. In time, science could grow toward a paradigm of
truly trans-disciplinary scope. Such a paradigm would be intensely
fraught with meaning. In light of it, in the words of philosopher
of science, Errol Harris, scientists could attempt to:

- account for the wholeness of the universe—a single, indivisible wholeness of distinguishable but inseparably related parts;
- furnish the principle of organization universal to the system, a principle immanent within all the parts of the universe each of which expresses and exemplifies it;
- provide the hierarchical scale of differentiation that stratifies all the parts in a progression of levels of emergent complexity, so that each successive part expresses and manifests the principle more fully and adequately than its precedecessors; and
- exhibit a complex network of interdependence where all elements are reciprocally adjusted in structure and function to one another.[6]

Regardless of the precise trajectory that the development of science would follow in the near future, in evolving toward the transdisciplinary dimension scientific theories will suggest radical changes in our habitual picture of the world. A coherent, systematically self-organizing universe is very different from the clockwork cosmos of Newtonian physics. The component entities that populate a self-creative universe are not mechanisms, or mechanistic entities, obeying the laws of classical mechanics with Laplacian determinism. Rather, these entities are dynamic, stochastically interacting and self-constituting "open systems." They are not reducible to the sum of their parts, but have emergent properties due to the complexity of their internal structures and external relations. Their interactions cannot be analyzed to simple and unidirectional chains of causes and effects: with multiple feedbacks and multicausal chains, all effects feed back ultimately to their causes. Their parts or components are mutually constitutive; they are neither replaceable nor interchangeable. Moreover the more complex systems cannot be modelled as mechanistic aggregates: they have dynamic processes that make them intrinsically self-correcting and goal-seeking. And their boundaries are fuzzy: where one system ends and another—or its environment—begins is more a matter of methodological convenience than a condition attributable to the systems themselves.

By contrast, the view that is still dominant in society is that human beings are categorically distinct from each other and from their environment: they compete with one another for survival; they manipulate their environment following the blueprints of physical, social and environmental engineering; they exploit nature for their own ends; and, in the last count, they seek only their own self-centered values and goals.

At a time when our societies transit to an interacting and interdependent web of technology, finance, production, consumption, and even culture and leisure, the current, not scientific but only "scientistic" image of the world must be replaced by a more meaningful and encompassing one. This is where the currently neglected responsibility of scientists comes in.

If the public's perception of the world picture projected by science lags behind the world picture suggested by front-line research, it is because the latter is not effectively diffused, nor indeed properly articulated. The blame cannot be unilaterally assigned to science education, nor to the publication media. The problem originates with the scientists who pursue leading edge research. Reports on new findings are generally couched in quantitative language that is unintelligible to those outside the given fields. In many fields, by the time popularized accounts are written, the accounts themselves refer to semi-obsolete facts; front-line research has moved on. Consequently in regard to the world picture of science, the up-to-date is unintelligible, and the intelligible is obsolete. A way must be found to shortcut the process whereby intelligible information is diffused regarding the meanings properly associated with up-to-date findings.

The problem is not insoluble. There has been a felicitous tendency lately on the part of both scientists and science journal editors to communicate the significance of technical reports to a broader public. When important findings are reported, whether in the theoretical or in the experimental domain, both *Science* and *Nature* often publish editorial assessments in a more popular language. *Scientific American* in the US and *New Scientist* in the UK, key most of their output to a wider readership, and some new international journals, such as *The Journal of Scientific Exploration* and *Science Spectra*, are specifically oriented toward an interdisciplinary audience. A few leading-edge scientists, including Stephen Hawking, Ilya Prigogine and Paul Davies, have produced widely read full-length books to communicate the concepts suggested by their theories. Other investigators have made a point of outlining the broader significance of their findings for the general scientifically interested public. For example, Bernard Haisch, Alfonso Rueda and Harold Puthoff published their remarkable finding regarding inertia as a zero-point field Lorentz force in the physics journal *Physical Review* A, but followed up their technical exposition with a highly intelligible narrative account of its implications for our understanding of matter and universe in *The Sciences*. [7]

Nevertheless, there is still a lag between the concepts evolved in leading edge workshops and laboratories, and the understanding of the meanings suggested by those concepts in the mind of the public. The lag is the most vexing when it occurs in science education courses aimed at the non-science major. This touches the heart of the problem. It is the non-scientist who needs to be better informed of front-line advances and their entailments in the sciences, for it is he or she who will have to make the critical decisions that could affect present and future generations. Such decisions are subtly influenced by the vision the decision-makers entertain of man and universe, and this vision, in turn, is not-so-subtly influenced by what they understand of science.

Catching up with the lag between the world picture implied by leading edge discoveries in science, and that which dominates the mind of the public at large, is not a superficial matter readily resolved by measures such as more care in choosing and editing textbooks. It goes deeper than that: it involves the very source of the science-related knowledge that circulates in society. Here the knowledge-creating scientists themselves are implicated. They must remember that science is not only a source of technology; it is also, and above all, a source of meaning.

Recently Karl Pribram pointed out that the term 'narrate' is closely linked to the Latin '*gnarus*' which in turn is kin to '*gnoscere*', to know. Thus narration is a form of knowing, whereas in 'science,' '*scire*' is kin to '*scindere*', to cut. Thanks to the analytical tools furnished by mathematics, twentieth century science has been eminently successful in its pursuit of '*scire*': "cutting." However, mathematical formulations by themselves are cognitively incomplete. The narrative aspects of science, the concepts and meanings to which the computations point, have been neglected, sometimes deliberately as in the Copenhagen interpretation of quantum mechanics. According to Pribram this neglect has led to a cover-up of many anomalies and lacunae, and has produced considerable malaise in some scientists. Einstein, Dirac, Bohm and Bell have all attempted to understand their formulations in physics; von Bertalanffy and Koestler in biology, and Piaget and Maslow in psychology. But, for the most part, the received wisdom in the classroom has emphasized the elegance of what has been achieved, often with the advice that any attempt at further understanding would simply confuse.[8]

Nevertheless, deeper understanding, though entailing the risk of confusion, is not an expendable component in the production of scientific knowledge. Scientists need to shoulder the responsibility for articulating the meaning of their findings, and not only the mathematics that validate them. It is no longer enough to state the formulas, and the observations and experimental protocols that have led to them; the meaning of the formulas regarding our understanding of the phenomena also need to be articulated. The responsibility of scientists includes the provision of an intelligible narrative whenever the findings concern new and different entities and processes, or new and different relations between familiar ones.

This alone, however, would not satisfactorily solve the problem. A mere commitment on the part of specialized scientists to articulate the wider meaning and significance of their findings would most likely lead to a bewildering array of half-thought-out ideas and more-or-less naive presumptions. Handing the problem to philosophers specifically trained in the elaboration of meaning might offer a better alternative; yet few philosophers are prepared to comprehend the full import of findings in the various, and sometimes highly esoteric, domains of leading edge science. A high level of specialization to one or another field of investigation, whether by philosophers of science or by the scientists themselves, always entails the danger of arbitrary *sui generis* interpretations, exacerbating the gap between the languages used in the different fields of scientific inquiry.

It would be wiser, and *ceteris paribus* preferable, to get the editorial boards of prestigious science journals to undertake a narrative assessment of the published findings in consultation with the authors. Alternatively they could convene a special panel for this purpose. These steps would add to the editorial workload, but would also ensure that the published materials are meaningful, besides being technically acceptable.

Obviously, not every finding published even in the most reputable science journal is likely to merit the effort of articulating its meaning; some may entail but a minute modification of existing knowledge, while others, even if more radical, may have a predominantly hypothetical character. A first task must thus be to determine whether a given finding merits or requires narrative interpretation.

Such a determination could be attempted by the editorial boards of science journals and book and monograph series, or by the panels convened by them for this purpose. Interpretation requirements could be established in reference to some consensual scale, measuring (i) *innovativeness*, and (ii) *validity* in regard to the reported findings. Should both these ratings reach or surpass a given value, the editors could require the author(s) to consult with them or their panels in view of producing a narrative that would articulate:

(a) *how the findings relate to received knowledge on the given field(s),* and

(b) *what the findings imply in regard to our interaction with the pertinent phenomena.*

In some natural science fields (theoretical physics, astronomy, paleontology, etc.) the latter may involve no more than techniques of observation and experimentation, but in others, especially in the ecological, human and social sciences, it could have major consequences for human life and well being.

Attaching narratives to scientific research reports need not hold up the publication of the pertinent papers; it is sufficient that the narrative assessments should follow within a reasonable interval of time. Such assessments must be non-dogmatic. The editors need to make clear that they are not to be attributed the status of definitive pronouncements, but should be seen merely as an initial attempt to "unpack" the meaning and state the implications of the findings.

Using the time-honored formula of reviews, discussions, and comments by those whose findings are reviewed and discussed, one or more issues of a journal or other publication could achieve an incremental clarification of the issues—or, at the very least, of the problems connected with the clarification of the issues.

The responsibility of the reporting scientists would generally end at this point; unless they specifically wished to do so, they would not need to go beyond collaborating in articulating the implications entailed by their findings. The responsibility of science writers and editors would extend further, however. The editors of science journals and of book and monograph series would need periodically to review the various discussions and debates and select the most important narratives for further, and conceivably far wider, diffusion. By publishing annual or semi-annual collections or summaries of the most significant items, they could enable science writers, textbook authors and editors to produce articles and book-length monographs for the general public. Also up-to-date and explicitly meaningful textbooks could result for a variety of science education offerings.

With a due acceptance of responsibility for meaning on the part of the science community, effective efforts could get under way to combine intelligibility with up-to-dateness, thereby diffusing intelligible, accessible knowledge for the current and next generation of students and concerned lay people.

Notes and References

1. Katsuhiko Yazaki, "Going Beyond Boundaries for Our Future Generations," in *Thinking about Future Generations*. The Institute for the Integrated Study of Future Generations, Kyoto, 1994.
2. Ervin Laszlo, *The Choice: Evolution or Extinction*. Tarcher/Putnam, New York 1993.
3. Information on the Club of Budapest and its programs may be obtained by writing to the Secretariat at Szentharomsagstr 6, 1014 Budapest, Hungary.
4. Ervin Laszlo, *The Interconnected Universe*. World Scentific, Singapore, London and New York, 1995.
5. Eric Chaisson, "The Emerging Life Era", in *Thinking about Future Generations.*, op. cit.
6. Errol E. Harris, *Cosmos and Anthropos*, Humanities Press, New York 1991.
7. cf. Bernhard Haisch, Alfonso Rueda and H.E. Puthoff, "Inertia as a zero-point-field Lorentz force," PhysicalReview A, 49.2 (February 1994); Bernhard Haisch, Alfonso Rueda and Harold E. Puthoff, "Beyond E = mc²" *The Sciences*, November/December 1994.
8. Karl Pribram, Afterword to *The Interconnected Universe*, op. cit.

4

Toward an Empathic Science: The Hidden Subtext for Fundamental Educational Change

Riane Eisler

The traditional approach to science education focuses on discrete bits of data or "facts" or at best on separately presented scientific theories. This approach is far too narrow, particularly when "facts" keep changing due to the exponential pace of scientific discoveries and there are intermittent changes in scientific paradigms. It makes even less sense from the perspective of the enormous impact science has on our lives—when we consider that science and technology can be used to vastly improve our lives or destroy us.

What is urgently needed is a much broader approach to science education: one that teaches us from early childhood on to think of science—how it was constructed, who constructed it, and how and by whom it now needs to be reconstructed—in its social and ideological context. Fortunately there is already movement in this direction. But this movement can be accelerated if we leave behind the conventional categories that today frame much of the discourse about the struggle for our future: categories such as modern versus traditional, right versus left, and religious versus secular.

Science, Domination and Partnership

Many people blame modern science and technology—particularly Cartesian rationalism and Newtonian science—for our escalating social, economic, and ecological crises. But if we look at the world as it was before the 18th-century Enlightenment, if we go back to the religious Middle Ages, we find despotic repression, chronic warfare, brutal sex, legally-sanctioned family violence, and Church-launched witch hunts, witch burnings, and other horrible public displays of the infliction of pain as the perogative of those in power. In addition, today's decried exploitation and domination of nature was not only commonplace for much of recorded history, but was often idealized.

For instance, in the Babylonian Enuma Elish, we read that the god Marduk created the world by dismembering the body of the goddess Tiamat, and how through this man gained dominance over the "chaotic" powers of nature and woman—both symbolized by Tiamat. Later, in the Judaeo-Christian Bible, in Genesis 1:28, we read that when God created humans in his image, he gave man dominance "over every living thing that moveth upon the earth."

So clearly the environmental problems we face today are not the result of science and technology per se—but of science and technology informed by a well-entrenched ideology idealizing "man's conquest of nature." Similarly, contemporary scientific dogmas that "explain" the domination of one race or ethnic group over another are merely updates of earlier religious and philosophical dogmas. Neither is the use of such dogmas to rationalize male dominance a modern invention—as evidenced by Aristotle's pronouncement that slaves and women were meant to be dominated, based on the "fact" that otherwise they would not have been born slaves or women rather than free men.

What *is* new is the unprecedented power of modern science and technology, and what is also new is the growing awareness that a fundamental reexamination of science education is urgent.

The perspective for this reexamination proposal derives from two decades of multi-disciplinary study of human society drawing from a much larger database than is customary. This database encompasses the whole of our history, including prehistory, and the whole of humanity, both its female and male halves.

This systems analysis makes it possible to see patterns that were not visible before. To briefly summarize, I found that underlying the great surface diversity of human societies are two basic possibilities for social and ideological organization: the dominator and the partnership models. I found that the configuration characteristic of each model transcends the conventional classifications of societies by location, period, ideology, and level of technological development. I also found strong evidence that the original direction of cultural evolution was more in a partnership direction, until there was, during a prehistoric period of great disequilibrium, a shift to the dominator model—and that underlying the many currents and crosscurrents in our time of disequilibrium is the struggle between a mounting partnership resurgence and dominator systems resistance. Moreover, and this is of particular relevance to the deconstruction and reconstruction of science in ways appropriate for a partnership rather than dominator future, I found that how the roles and relations of the two halves of humanity—women and men—are defined critically impacts the social construction of a society's beliefs and institutions, including the social construction of scientific institutions and beliefs.

In societies that orient primarily to the dominator model, "real" masculinity is characteristically equated with domination and conquest. Hence an integral feature of male socialization is the idealization of male violence and domination as heroic (be it through epics, childhood games, or histories focusing on who won and lost wars), along with the early inculcation of contempt for the "feminine." In addition, both control over economic resources and policy-making (be it religious, political, scientific, educational, etc.) is in these societies deemed appropriate only for men.

By contrast, in societies orienting more to the partnership model, what we have been conditioned to consider stereotypically feminine values, emotions, and activities—such as care-taking, empathy, and

nonviolence—are not devalued. Both women and men are social-
ized to give high priority to them, and both women and men have
important roles in the administration of economic resources and
in decision-making, be it in family, religious, political, economic,
educational, etc., institutions.[1] (Please see endnote for some examples).

What all this points to is that it is not possible, as many religious sages and philosophers have urged, to move to a society
governed by a more stereotypically "feminine" ethos of caring,
nonviolence, and compassion without fundamental changes
in the stereotypical roles and relations of the female and male
halves of humanity.

What it also points to is that if we are serious about a science
education that can help our children, and future generations, use
science and technology in more humanistic ways, we need to pay
close attention to what the underlying dynamics governing the
development and use of science have been and can be.

The Deconstruction of Science

We hear a great deal of rhetoric today attacking modern Western
science in the name of religion. But ironically, the development of
modern Western science came out of religious institutions. Indeed, the entire academy as we know it today was to a large extent
an outgrowth of the Christian clerical culture of the Middle Ages.
As the historian of science David Noble notes in his book *A World
Without Women: The Christian Clerical Culture of Western Science*,[4]
all the practices and institutions of higher learning were dominated by Christian clerics, whose culture from the late medieval
period on militated against the inclusion of women—as well as
"feminine" values—in scientific enterprise.

"However some historians might retrospectively characterize
Western science as a secular enterprise," he writes, "it was always
in essence a religious calling, more a continuation of than a
departure from Christian tradition...At the outset, of course, the

culture of science was the culture of the ecclesiastical academy and, hence, a world without women... Western science thus first took root in an exclusively male—and celibate, homosocial, and misogynist—culture, all the more so because a great many of its early practitioners belonged also to the ascetic mendicant orders." (Noble 1992, 163).[4]

Noble then points out that "several habits and characteristics of modern science have often been noted: the strict separation of subject and object, the priority of the objective over the subjective, the depersonalized and seemingly disembodied discourse, the elevation of the abstract over the concrete, the asocial self-identity of the scientist, the total commitment to the calling, the fundamental incompatibility between scientific career and family life, and, of course, the alienation from and dread of women with which this study opened." (Noble 1992, 282).[4]

I agree with Noble that many of the contemporary "characteristics and habits" of science betray its clerical legacy, including the "long-standing clerical effort to subdue the feminine in society and nature, in order to effect man's recovery from the Fall—'as if he had never sinned.'" (Noble 1992, 282),[4] However, I do not believe that we can attribute these characteristics solely to the ascent of Western clerical culture. Rather, it is rooted in the shift that occurred during our prehistory from an earlier cultural evolution orienting more to a partnership rather than dominator model—a time when there was more equal valuing of both halves of humanity, as well as of the characteristics we today stereotypically associate with femininity, including being an empathic rather than "objective" observer.

The deconstructionist/postmodern recognition that in reality there is no such thing as objectivity, that we are all products of our cultures and our life histories, and that therefore much of what we consider "reality" is a matter of interpretation is certainly an important development. However to merely substitute "detached irony" (the battle cry of postmodern discourse) for "detached objectivity" (the battle cry of modern science) is hardly a new approach to scientific inquiry. Actually, it comes to the same. For what is lacking in both is feeling, or more specifically, empathy—a lack that has all too often made modern science a tool for maintaining the massive inequities and imbalances inherent in a dominator status quo and that, at our level of technological development, threatens the very survival of life on this planet.

The Humanistic Reconstruction of Science

What can we as educators, as scientists, as parents and grandparents, do to change the education of both girls and boys for a world where science will be associated with empathy rather than detachment, with caring rather than control, where its major uses will no longer be for stereotypically "masculine" matters such as ever "better" armaments and ever more "effective" wars, but rather for the "women's work" of feeding all our children, caring for all our people's health, developing all our people's unique human potentials—in short, a science informed by empathy and other stereotypically "feminine" values?

I believe that there is a great deal we can do. But it requires leaving behind many of our preconceptions about science, education, and even beyond this, about what it means to be a woman or a man.

For one thing, we need to teach children to recognize that science has not been objective in the sense of neutral or unbiased, that much of its methodology and focus have been shaped by normative ideals of masculinity appropriate for precisely the kind of society we are today struggling to leave behind. We need to help them recognize the social and ideological context of the scientific practices and institutions we have inherited, to identify underlying patterns such as the dominator and partnership configurations, and to fundamentally re-examine what the uses of science and technology should be.

Children need to learn systems thinking—for example, that the so-called private and public spheres are inextricably interconnected and that whether the relations between the two halves of humanity are structured as rankings of domination or equal partnerships is a basic model for all relations. They also need a much more gender-holistic curriculum—one that no longer focuses primarily on the ideas and activities of half of humanity. They need a curriculum that gives far less emphasis to men's struggles for control through wars and revolutions and far more emphasis to women's and men's struggles in both the so-called private and public spheres against violence and for greater equity.

They need role models, both female and male, of people and groups dedicated to humanistic social change. And they need to be educated to understand that all human pursuits are guided by systems of values, that there is no such thing as a values-neutral science, and that the real question is *what kinds of values* inform both our personal and public choices—including our choices regarding science and technology.

In this regard, however, we need to emphasize that the key questions regarding values are not questions of religious versus secular, right versus left, or traditional versus modern or postmodern. Rather, the key questions for our future, as David Loye shows, are whether the kinds of values that inform social, scientific, and technological decision-making are congruent with a partnership or dominator model.

For what we have today as a result of a 5,000-year dominator era are technologies of destruction that can snuff out all life on this planet—a power that was, significantly, in more religious times, attributed to a supreme father deity. Moreover, man's fabled "conquest of nature" now has at its disposal equally lethal technologies: for example, clear-cutting technologies that can denude our rainforests (the lungs of our planet), depriving us of the essentials that sustain life in all world regions; chemical technologies, such as pesticides that pollute our air, land, and water in the name of progress (again destructive powers once attributed only to punitive supernatural entities); and even the potential power to wrest from women the birthing of life through its creation in laboratories (thus potentially the capacity to actually create a world without women).

Small wonder that there is today a reaction against science and technology. But again, the underlying problem is not science and technology per se, but the guiding ethos informing science and technology and with these, the hidden subtext of gender that has been such a strong thread holding together traditions of domination and violence. It is this then that needs to be addressed if we are to effectively educate our children for a world where science and technology can be used in humanistic rather than anti-human ways.

Notes and References

1. For example, the dominator configuration of rigid male dominance, strongman rule in both the family and state, a high level of institutionalized violence (ranging from wife and child beating to warfare), and the idealization of what is considered "masculine" can be found in tribal societies such as the low technology 19th Masai, highly technologically developed societies such as Nazi Germany, communist societies such as Stalin's Soviet Union, mideastern theocracies such as Khomeini's Iran, and oriental warrior societies such as the Samurai of Japan. Moving to the partnership side of the spectrum, we find tribal societies such as the Tiruray of the Philippines and the Bambuti of the Congo, highly technologically developed societies such as today's Scandinavian block nations (in which *not* coincidentally both a rise in the status of women and the social funding of "women's work" such as feeding children, caring for people's health, and environmental housekeeping occurred at the same time), and the prehistoric societies in both Western and Eastern civilizations described in works such as my book *The Chalice and the Blade: Our History, Our Future*[2] and the recently published *The Chalice and the Blade in Chinese Culture,*[3] a compendium of essays by archaeologists, sociologists, political scientists, and other scholars from the Chinese Academy of Social Sciences in Beijing.

2. Eisler, Riane, *The Chalice and The Blade: Our History, Our Future.* San Francisco Harper & Row, 1987.

3. Min, Jiayin, editor, *The Chalice and the Blade in Chinese Culture: Gender Relations and Social Models*. Beijing China Social Sciences Publishing House, 1995.

4. Noble, F. David, *A World Without Women: The Christian Clerical Culture of Western Science*. New York: Alfred A. Knopf, 1992.

5. Eisler, Riane, *Sacred Pleasure: Sex, Myth, and the Politics of the Body*. San Francisco: HarperSan Francisco, 1995.

6. Eisler, Riane, David Loye, and Kari Norgaard, *Women, Men, and the Global Quality of Life.*, Pacific Grove, California, Center for Partnership Studies, 1995.

7. Keller, Evelyn Fox, *A Feeling for the Organism: The Life and Work of Barbara McClintock*. San Francisco: W.H. Freeman, 1983.

8. Loye, David, *The River and The Star: The Lost Story of the Scientific Exploration of Goodness*. Work in progress.

5

Cosmological Education for Future Generations

Brian Swimme

In this challenge of thinking about future generations, I would like to offer some ideas having to do primarily with the children of future generations, and above all, the education of the children of future generations.

One source of inspiration for my reflections is the educational work of the past generations. I am always deeply impressed by the amount of energy earlier generations devoted to the task of teaching *cosmology*, by which I mean teaching the fundamental nature of the universe as well as the proper human orientation within the universe. We see something of this in the magnificent cave paintings from as far back as 20,000 years ago in southern Europe, when our ancestors crawled for days on their backs through the labyrinthine caves to gather in great underground vaults for their cosmic celebrations. Other artifacts of cosmological reflection reach back perhaps 40,000 years, and some anthropologists surmise that our human ancestors have been gathering in caves and teaching each other for as long as 300,000 years.

Modern humanity seems to be the first culture to break with this primordial tradition of celebrating the mysteries of the universe. Modern industrial society does it differently. Questions of ultimate meaning and value are dealt with not in caves or on the open plains, but in the churches, mosques, and temples. Here each weekend billions of humans gather to reflect on their relationship with the God or the Divine Ground. In all these millions of weekly religious ceremonies, so essential to the health and spirituality of humanity as a whole, one will find such a diversity of religious celebrations, but only rarely will one find serious contemplation of these primal human questions *within the context of the actual universe*, a universe of stars, topsoil, amphibians, and wetlands.

The result is that, within our religions, when we do ponder the deep questions of meaning in the universe, we do so in a context fixed in the time when the classical *scriptures* achieved their written form. We do not worship or contemplate in the context of the universe as we have come to know it over these last centuries, a context that includes the species diversity of the Appalachian mountains, the million year development of the Chinese ecosystems, the intricate processes of the human genome, or anything else that is both specific and true concerning the Earth and universe. All of that—the Earth and universe as they are and as they actually function—is regarded as "science", something separate from questions of meaning and value which religions deal with.

Modern humans, instead of gathering in the caves or cathedrals to dance to poetry and music as a way of learning their place in the universe, sit in classrooms and study science. Certainly such education in the sciences is fundamental for the survival of humanity. The challenges that beset us today will grow ever more fierce for our children and their children, and we require the best science and technology we are capable of. But nowhere in science education—not in Europe or Asia or the Americas—is the fundamental role and meaning of the human and the universe treated in any significant manner. The ruling assumption is that science is concerned with facts; whereas meaning and purpose and value are the domain of religion.

The tragedy here is that our religions would remain true to their essence if they were to think and work within the larger context of the universe. It would not mean shrinking away from their central truths. On the contrary, expressed within the context of the dynamics of the developing universe, the spiritual truths of the planet's religions would find a far vaster and more profound form. The recasting would not be a compromise nor a belittlement; it would be a surprising and creative fulfillment, one whose significance goes beyond today's most optimistic evaluations of the value of religion.

But if humans, in order to become fully human, truly do need to ponder the universe to discover their place in nature, and if this 300,000 year tradition is rooted in the requirements of our genetic make-up, then we will find our way to ideas concerning the proper human role in the universe one way or another. And if the institutions of education and religion have, for whatever well-defended reasons, decided to abdicate that role, someone somewhere else is going to step forward and provide it.

Where are we initiated into the universe? To answer we need to reflect on what our children experience over and over again, at night, in a setting similar to those children in the past who gathered in the caves and listened to the chant of the elders. If we think in terms of pure quantities of time the answer is immediate: the cave has been replaced with the television room and the chant with the advertisement.

What is the effect on our children? Before a child enters first grade science class, and before entering in any real way into our religious ceremonies, a child will have soaked in 30,000 advertisements. The time our teenagers spend absorbing ads is more than their total stay in high school. What we need to confront is the power of the advertiser to promulgate a world-view that is based upon dissatisfaction and craving, as is clear from one of the cliches of modern advertising: "An ad's job is to make them unhappy with what they have."

Advertisements are where our children receive the cosmology of the modern age. Perhaps the more recalcitrant children will require 100,000 ads before they cave in and accept consumerism's basic world-view. But eventually we all get the message. It's a simple cosmology, told with great effect, and delivered a billion times each day to nearly everyone in the planetary reach of the ad: *humans exist to work at jobs, to earn money, to get stuff.* That's paradise. And the meaning of the Earth? Implicit in every ad is the dogma that Earth is simply premanufactured consumer stuff.

If we are to learn from our ancestors we must devote ourselves to replacing the propaganda of the ad with a new cosmology, a new cosmology for a new millennium, one that is rooted in the understanding of contemporary science and nourished by the deepest wisdom of our philosophical and religious traditions. To indicate what I have in mind I would like to speculate on how we might introduce the children of future generations to the Sun.

The science I would begin with is our discovery of the nature of the Sun's energy. With the physics developed in our own century, we have learned something truly amazing: the Sun, each second, transforms four million tons of itself into light. Each second a huge chunk of the Sun vanishes into radiant energy that soars away in all directions. In our own experience we have perhaps watched candles burn down, or have seen wood consumed by flames and leaving behind only ashes, but nothing in all our human experience compares to this preternatural blaze that engulfs oceans of matter each day.

Here is yet another gateway through which a new cosmological imagination approaches a synthesis of science and religion. In the case of the Sun's energy, we are presented with a new understanding of the cosmological meaning of sacrifice. The Sun is, with each second, giving itself over to become energy that we, with every meal, partake of. We so rarely reflect on this basic truth from biology, and yet its spiritual significance is supreme. The Sun converts itself into a flow of energy that photosynthesis changes into plants that are consumed by animals. So for four million years, humans have been feasting off the Sun's energy stored in the form of wheat or maize or reindeer as each day the Sun dies as Sun and is reborn as the vitality of Earth. And those solar flares are in fact the very power of the vast human enterprise. And every child of ours needs to learn the simple truth—she is the energy of the Sun. And we adults should organize things so her face shines with the same radiant joy.

During the modern period when materialism came to dominate, such a suggestion as this last would be rejected as "mere poetry." It would be labelled a "projection" and ignored. We simply did not know that the actual energy coursing through our respiratory and nervous systems was bestowed upon us by the Sun and that our own vitality is a natural evolutionary development of the Sun's vitality. So instead of introducing our young to the Sun we cut them off from the Sun. That is, instead of awakening this primordial relationship that would shine on the child's face with the radiance of the Sun, we unknowingly and tragically snuffed it out. They were left only with the option of following our own convictions, that the universe was a collection of dead objects, and so it went from generation to generation throughout the modern world.

In the cosmology of the new millennium, the Sun's extravagant bestowal of energy can be regarded as a spectacular manifestation of an underlying impulse pervading the universe. In the star this impulse reveals itself in the ongoing give-away of energy. In the human heart it is felt as the urge to devote one's life to the well-being of the larger community.

In a culture where cosmology is living, children are taught by the Sun and Moon, by the rainfall and starlight, by the salmon run and the periwinkle's hideout. It has been so long since we moderns have lived in such a world, it is difficult to picture, but we can just now begin to imagine what it might be like for our children, or for our children's children.

They will wake up a few moments before dawn and go out into the gray light. As they're yawning away the last of their sleep, and as the Earth slowly rotates back into the great cone of light from the Sun, they will hear the story of the Sun's gift. How five billion years ago the hydrogen atoms created at the birth of the universe came together to form our great Sun that now pours out this same primordial energy from the beginning of time. How some of this sunlight is gathered up by the Earth to swim in the oceans and to sing in the forests. And how some of this has been drawn into the human venture, so that they themselves are able to stand there, they are able to yawn, they are able to think only because coursing through their blood lines are molecules energized by the Sun.

And then they will hear the simple truth about the necessity of such a bestowal. If we burn brightly today, it is only because this same energy was burning brightly as the Sun a month ago. Even as we take a single breath, our energy dissipates and we need to be replenished all over again by the Sun's gift of fire. If the Sun were suddenly to stop transforming itself into energy, all the plants would die as the Earth's temperature plummeted hundreds of degrees below zero. In our veins and flesh, all the heat-giving molecules would go cold without replenishment as we and everything else became hard as frozen stone.

The Sun's story will find its climax in a story from the human family of those men and women whose lives manifested the same generosity, and whose sacrifice enabled others to reach fulfillment. If through the ages the various cultures have admired such people who poured out their creative energies so that others might live, we were only intuitively recognizing that such humans were true to the nature of the energy that filled them.

Human generosity is possible only because at the center of the solar system a magnificent stellar generosity pours forth free energy day and night without stop and without complaint and without the slightest hesitation. This is the way of the universe. This is the way of life. And this is the way in which each of us joins this cosmological lineage when we accept the Sun's gift of energy and transform it into creative action that will enable the community to flourish.

Of course, over the years, as the Sun's story is repeated even in its various forms, there will be a good deal of repetition, and the listeners will sometimes be bored and distracted. This is to be expected. This is not entertainment but education. And moral education in particular rests upon holding in mind, over long periods of time, the magnificent achievements within the universe.

By reminding ourselves of the possibilities of true greatness and true nobility of spirit, we excite the energies necessary to achieve our true fulfillment. Then the challenge of moral and spiritual achievement is not something dealt with for an hour on the weekend. Then the task of transformation is the way we start each day as we remind ourselves of the revelation that is the Sun.

Through repetition, and through years of deepening, our children or our children's children will be provided a way to escape the lures of so much deceit, greed, hatred, and self-doubt, for they will begin each morning and live each day inside the simple truth: a gorgeous living Earth drifts light as a feather around the great roaring generosity of the Sun.

6

Tolstoy, Napoleon and Gompers

Freeman Dyson

Samuel Gompers

One of my favorite monuments is a statue of Samuel Gompers not far from the Alamo in San Antonio, Texas. Under the statue is a quote from one of Gompers's speeches:

> *What does labor want?*
> *We want more school houses and less jails,*
> *More books and less guns,*
> *More learning and less vice,*
> *More leisure and less greed,*
> *More justice and less revenge,*
> *We want more opportunities to cultivate our better nature.*

Underneath this quote, it says:

To the Doers
Dedicated September 6, 1982
Sculptor Bette Jean Alden

I do not know how many of the thousands of tourists, who come every year to San Antonio to pay their respects to Davy Crockett and the other heroes of the Alamo, take a few moments on the way to show respect also to Samuel Gompers. I hope some of them find, as I did, the peaceful words of Gompers a fitting antidote to the cult of military madness symbolized by the Alamo. It comes as a refreshing surprise to hear, so close to the heart of Texas, so close to the shrine of patriotic pride, a quiet voice of

53

reason. Samuel Gompers was the founder and first president of the American Federation of Labor. He was largely responsible for the fact that the American labor movement broke away from the European movement dominated by the ideology of Karl Marx. The European labor leaders dreamed of a proletarian revolution. Gompers understood that American working people were not interested in revolution and dreamed mostly of high wages and economic security. He understood that American working people were also interested in education and self-improvement.

I chose the words of Gompers to introduce this essay because he was the champion of pragmatism against ideology. It is an ironic fact of history, that now, seventy years after his death, the ideas of Gompers have triumphed in Europe and failed in the United States. In Gompers' lifetime the revolutionary ideologues of Europe led their unions into disaster after disaster, the worst disaster being the dictatorship of the proletariat in Russia in 1917. Meanwhile, Gompers established in America the tradition of practical bargaining between labor and management which led to an era of growth and prosperity for the unions. Then, after the devastations of World War II, the positions on the two sides of the Atlantic Ocean were gradually reversed. Western Europe recognized the wisdom of Gompers and rebuilt its societies on a non-ideological basis, the unions sharing power and responsibility for economic decisions. Meanwhile, the United States forgot Gompers and embraced an ideology of doctrinaire free-market capitalism. In America the unions dwindled, while Gompers's dreams, more books and less guns, more leisure and less greed, more school-houses and less jails, were tacitly abandoned. In a society without social justice and with a free-market ideology, guns, greed and jails are bound to win.

This essay is not supposed to be about social justice. It is supposed to be about education and science. But you cannot tell which kind of scientific education is good and which is bad without paying some attention to social justice. When I was a student in England fifty years ago, one of my teachers was the great mathematician

G. H. Hardy, who wrote a little book, *A Mathematician's Apology*, explaining to the general public what mathematicians do. Hardy proudly proclaimed that his life had been devoted to the creation of totally useless works of abstract art, without any possible practical application. He had strong views about science, which he summarized in a famous sentence. He wrote, "A science is said to be useful if its development tends to accentuate the existing inequalities in the distribution of wealth, or more directly promotes the destruction of human life". He wrote these words while a particularly destructive war was raging around him. Still, the Hardy view of science has some merit even in peace-time. Many of the useful sciences which are now racing ahead most rapidly, replacing human workers in factories and offices with machines, making stock-holders richer and workers poorer, are indeed tending to accentuate the existing inequalities in the distribution of wealth. And the technologies of lethal violence continue to be as profitable today as they were in Hardy's time. It is no wonder that a majority of our children have turned away from science. The more you tell them that science is useful, the more they are turned off. They know what useful means. To bring them back to science, we need to give them the kind of education that Gompers dreamed of, education as an opportunity to cultivate their better nature, in a society with more school-houses and less jails, more books and less guns, more leisure and less greed.

Napoleon and Tolstoy

In a society that has forgotten Gompers, the question remains: what kind of education should we provide for our children? Two other historical figures, Napoleon and Tolstoy, may help us to answer this question. They were both passionately involved with education. And their views about the nature and purpose of education were as strongly opposed as their views about war and peace.

Napoleon understood that public education was important for the modernization of France and the staffing of an Imperial government. He created a system of education in France for these purposes. He imposed similar systems on the countries of Europe that he conquered. The basic pattern of European public education has remained for two hundred years as Napoleon established it. The basic pattern is centralization, standardization and rigorous testing by examination. Education was organized rigidly from the top down. The state authorities, beginning with Napoleon himself, decided what was to be taught, when, and how. On the whole, the Napoleonic system has done successfully the job for which it was designed. In the countries that adopted it, it produced populations with high levels of literacy and technical competence. It has also acted as a social equalizer, giving talented children from all classes of society an opportunity to become leaders. Napoleon stated explicitly that his aim was to open careers to talents. In many ways, the rigidity of the Napoleonic system works to the advantage of an underprivileged child with talent. The child does not have the option of refusing to join the educated elite. Chiara Nappi, a physicist colleague of mine in Princeton who was educated by the Napoleonic system in Italy and serves on the Princeton school board, has argued convincingly that a small dose of Napoleonic rigidity could be helpful to the underprivileged children of America. She knows from her own experience that a little Napoleonic rigidity can be especially helpful to girls with a talent for science. Girls in Italy do not have the option of dropping out of science. Our American system of education gives children of all ages too many opportunities to drop out. We still have something to learn from Napoleon. But we also have something to learn from Tolstoy.

Tolstoy devoted three years of his life to education. He did not work like Napoleon from the top down. He worked from the bottom up. He organized a school for the children of the peasants on his estate at Yasnaya Polyana. He taught the children himself. He wrote an extensive account of what he had taught them and what he had learned from them. His purpose was to find out, by personal observation at the individual level, how teaching could and should be done.

Before starting his own school, Tolstoy traveled extensively around Europe and observed the Napoleonic system of public education in action. What he saw convinced him that the Napoleonic system was a total failure. Both in Russia and in other European countries, he wrote, "All that the greater part of the people carry away from school is a horror of schooling". He describes the typical child as he saw it in the class-room, "*A* weary, huddled creature, with an expression of fatigue, terror, and boredom, repeating with the lips alone the words of others in the language of others, a creature whose soul has hidden in its shell like a snail". He contrasts this dismal scene with "the same child at home or in the street, full of the joy of life and love of knowledge, with a smile in its eyes and on its lips, seeking instruction in all things as a joy, expressing its own thought clearly and often forcefully in its own language".

Unfortunately, Tolstoy had a short attention-span. For three years he gave his school and his pupils his passionate attention, and then he abandoned them. A year later, he wrote in a letter to a friend, "The children come to me in the evenings and bring with them memories of the teacher that used to be in me and is there no longer. Now I am a writer with all the strength of my soul". The children sadly reported to Vasily Morozov, who had been Tolstoy's assistant in the school, "*We* could not get on with Lev Nikolayevich as we used to". All that is left of the school is the memoir of Vasily Morozov and the descriptions that Tolstoy recorded in his journal. Tolstoy's descriptions are vivid and detailed. He was, after all, born to be a writer rather than a teacher. He describes how he taught creative writing to Syomka and Fedka, two of his favorite pupils. He begins by trying to force them to analyze the written text of a story by Gogol. The task of analyzing the language leaves the children puzzled and bored. Tolstoy remarks, "You, the teacher, are insisting on one side of understanding, but the pupil has no need whatsoever of what you want to explain to him". Finally, Tolstoy suggests to Fedka that he write a story himself. Tolstoy gives him the theme, a peasant family with a father who leaves home to become a soldier and then returns after six years in the army. Fedka writes the story, and Tolstoy analyzes the text. The process of textual analysis, which had been so sterile when applied to Gogol, comes alive and acquires meaning when it is applied to Fedka's own words. Fedka successfully defends his choice of words against Tolstoy's criticism. In a short time Fedka has become, not merely a gifted story-teller, but a conscious artist aware of his own technical mastery.

Anyone who wishes to learn more about Tolstoy's educational methods and experiences can find them, together with further references to the literature, in a long article by Michael Armstrong, "Tolstoy on Education: The Pedagogy of Freedom", in the summer 1983 issue of *Outlook*. *Outlook* was a quarterly magazine founded by David Hawkins, the philosopher of Los Alamos, and devoted to the education of young children. Unfortunately, like Tolstoy's school, it is now defunct. Its demise does not make its ideas and opinions any the less valuable in the world of today.

Napoleon and Tolstoy both proclaimed important truths about education. Napoleon proclaimed that the essence of education is discipline. Tolstoy proclaimed that the essence of education is freedom. These two doctrines seem to be diametrically opposed. How can both of them be true? They can both be true because Napoleon and Tolstoy were concerned with different kinds of children and different meanings of the word education. For Napoleon, education meant the training of an intellectual elite who would find careers in the institutions of a technically organized society. For Tolstoy, education meant opening the eyes of ordinary children to the wonders of the world around them. For Napoleon, education meant cramming docile children with specialized knowledge. For Tolstoy, education meant giving all children, docile or not, direct experience of creative activity in art and science and literature. In the modern world, even more than in Tolstoy's world, both kinds of education are necessary. We need to give every child a taste of the Napoleonic discipline that leads to useful skills and brilliant careers. And we need to give every child, especially the child who is turned off by Napoleonic discipline, a taste of the Tolstoyan freedom that leads to the growth of natural scientific curiosity and artistic expression. Only Napoleon can give us a cadre of technically trained experts to make our society economically competitive. Only Tolstoy can give us a population of culturally and scientifically literate citizens who may possess the wisdom to save our society from Napoleonic follies. Our systems of public education can use all the help they can get, both from Napoleon and from Tolstoy. We have a long way to go.

7

The Thirteenth Labor of Hercules

Dudley Herschbach

Education is what's left after all you've learned has been forgotten.
— Attributed to James B. Conant (among others)

In nearly 40 years of teaching, I've given many different courses in chemistry, graduate and undergraduate; advanced, intermediate, or introductory. In every course, at the outset I've always cited the saying quoted above. It defines the aim to be understanding rather than training, self-reliant thinking rather than conventional knowing. Since my own student days, I've cherished the "what's left" aspects of science and mathematics, which offer much that transcends any technical particulars. Especially appealing is the human adventure of intellectual exploration, replete with foibles and failures, but ultimately achieving wondrous insights. This I have sought to emphasize in my courses, not merely as seasoning for a hearty diet of chemistry, but to nurture perspectives akin to the liberal arts.

In pursuing this "liberal science" approach, I particularly enjoy the opportunity to indulge in whimsical fun while delivering a serious message. Such is my intent here. As in my freshman chemistry course, I discuss a fancied task that might have been asked of Hercules. What if that mighty hero, after completing his legendary twelve labors, had been required to weigh the earth's atmosphere? Perhaps he would have failed. If so, it would testify that even superhuman strength and courage cannot prevail when what is needed is a new intellectual concept. Perhaps he would have succeeded. If so, that would lead to consequences far more marvelous than his traditional twelve triumphs. Either way, this thirteenth labor takes on the character of a parable.

After sketching the parable, I consider labors posed by concerns about science literacy and public understanding of science. Optimistically, I suggest some ways scientists and science teachers might budge persistent, daunting attitudes that would have fazed Hercules.

From Water Pump to Gas Laws

Anyone who has had high school chemistry has encountered the gas laws; the mere mention elicits: $PV = nRT$! However, rarely is this accompanied by a cultural perspective. Lacking is any notion of how such prototypical concepts emerged, how widely applicable they are, or how they led to other developments. Stories that present science in a more humanistic mode can disarm fears, reveal a much broader context for nominally familiar concepts, and even induce students to relate the tales to others. My lecture on the gas laws concludes with remarks about Hercules, but it starts with a water pump that confounded Aristotle and Galileo. That part of the story I found in an excellent book, *On Understanding Science*,[1] published 50 years ago by James Conant. He advocated presenting science to "laymen through close study of relatively simple case histories," and his book provides several fine examples.

In much of the world, farmers and villagers still pump water by means of a simple suction device, in use since ancient times. Lifting a close-fitting piston in a vertical tube creates a rough vacuum. When the tube is immersed in a well or river, water rushes up the tube, like soda in a straw. Aristotle accounted for this by his famous dictum that "Nature abhors a vacuum." Probably that is still accepted as an explanation by many people today. However, such a pump will not lift water above a height of 34 feet. This empirical fact was known in Aristotle's day, as evident from artwork that depicts a series of pumps lifting water from a deep gorge, with human figures providing the scale. Aristotle said nothing about why a tall drink seems to quench Nature's abhorrence.

Two thousand years later, Galileo considered specifically the question why the pump would not lift water over 34 feet. He suggested that the pump ceases to function because a taller column of water would break of its own weight. That answer is also quite wrong. Contrary examples are familiar in waterfalls and fire hoses. (An intriguing detour into strength of materials may branch off here; a good source is *The New Science of Strong Materials*, by J.E. Gordon.[2])

The right idea was proposed by Torricelli, one of Galileo's students. I enjoy pointing out to my students that some of them are likewise destined to solve problems that have long stumped their teachers. Galileo knew that air had weight and had even devised a means to weigh it, but he did not connect this with the operation of a water pump. Torricelli realized that the weight of the air would force water to rise in the pump barrel. This concept implied that the observed limit of 34 feet represented the weight of water that the pressure of the air on the earth's surface could maintain.

To test his idea, Torricelli tried an experiment. For convenience, he used mercury, a liquid about 14 times heavier than water. If he was right, the atmospheric pressure should support a column of mercury only about 1/14 as high as that of water, or about 30 inches. His apparatus was simply a glass tube about three feet long, with one end sealed. He filled it with mercury, then inverted the tube in a bowl of mercury open to the atmosphere. In repeating this experiment for my classes, I'm always elated to see the mercury column in the tube drop to a height of about 30 inches above the level in the bowl. From weather reports, everybody knows about variations in atmospheric pressure, but few are aware that it is still measured by Torricelli's barometer, in essentially the same form he devised 350 years ago. Moreover, it is easy to demonstrate how vacuum pumps, evolved from the barometer, enabled measurements that established the gas laws.

The story thus far offers several morals. It illustrates well how a
maverick idea, tested by experiment, can overthrow long accepted
doctrines. The vacuum left between the top of the mercury column
and the sealed end of the glass tube refuted Aristotle's dictum. But
his venerable authority did not yield quietly. Many scholarly papers
in Torricelli's day tried in vain to save the old view by postulating
such things as invisible threads holding up the mercury. The story
also shows how a new conceptual paradigm gives rise to experimental techniques that further extend its scope. Above all, it exemplifies how profound insights may lurk in seemingly mundane
observations.

Weighty Atmosphere with Flighty Things

The proposed 13th labor of Hercules is easily accomplished
with the aid of Torricelli's barometer. From the pressure exerted
by 30 inches of mercury or 34 feet of water, the equivalent atmospheric pressure is determined to be about 15 pounds per square
inch. That amounts to about 30 million tons per square mile. To
obtain the total weight of the atmosphere, we need only multiply
by the estimated surface area of the earth, about 200 million square
miles. Thus, Hercules should have found that the earth's envelop-
ing ocean of air weighs about 6 billion megatons (6×10^{15} tons).

From this result, in homework labors students can readily derive
many other quantities of interest. For instance, it is easy to estimate
the total number of air molecules in the atmosphere. Using the
molecular weight of air (29 g/mol) and the famous Avogardro
number (6×10^{23} molecules/mole) with appropriate unit conver-
sion factors, we find that 6 billion megatons of air contains about
10^{44} molecules. From the gas law and the volume of our lungs, we
can estimate that in each breath we take in about 0.5 grams of air,
or 10^{22} molecules. Hence a breath is the same fraction of the whole
atmosphere ($10^{22}/10^{44}$) as a single molecule is of that breath ($1/10^{22}$). Nitrogen, the major component (80%) of air, is quite unre-
active; so most of the nitrogen molecules we breathe are quite

venerable, hundreds or even thousands of years old. Accordingly, if we assume the last breath of Aristotle or Galileo (or anyone else) has become uniformly spread through the atmosphere, it is quite likely that we inhale one nitrogen molecule from that gasp in each breath we take.

Macroscopic consequences of air pressure are likewise striking. Among such phenomena, all related to the fateful 34 feet of water, are how airplanes fly, sailboats sail, and curve balls curve. These are discussed in a charming book: *Rainbows, Curve Balls, and Other Wonders of the Natural World Explained* by Ira Flatow.[3] He describes how air flowing around a plane wing or boat sail of curved cross section or a spinning baseball is thinned out on one side relative to the other. That creates a pressure differential and thereby a force proportional to the surface area of the wing, sail, or ball.

This force can become superherculean. When fully laden with fuel and passengers, a 747 jumbo jet weighs about 350 tons. The surface area of its wings is roughly 3500 square feet. Atmospheric pressure of 15 lbs per square inch corresponds to about one ton per square foot. The massive plane therefore can be lifted aloft if there exists a pressure differential between the top and bottom of its wings of more than about 1/10 of atmospheric pressure, achieved by driving the wings through the air at sufficiently high speed. The same considerations apply to a gliding seagull, but the required pressure differential is only about 1/1000 of atmospheric pressure. Huge or puny, the aerodynamic force in its many variants, all akin to the barometer and water pump, exhibits the unifying power of Torricelli's idea.

Liberal Science and Science Literacy

The freshman chemistry course I've taught in recent years, with emphasis on "what's left," has been nicknamed by the students "Chem Zen." (In the catalogue it is Chemistry 10). Perhaps naively, I like to think that nickname endorses the liberal science approach. It is implemented not only with many parables but with poetry contests, unconventional labs and homework problems (e.g., treating metabolism of beer, healing of bones, flight of jumbo jets, restoration of Humpty-Dumpty...). A chief aim is to entice students to take ownership of scientific ideas, by applying them in familiar contexts that offer engaging perspectives.

As indicated, my own teaching has thus far been entirely in college courses intended for students of science. For such courses, emphasizing what I have termed liberal science may seem a distraction. However, I regard the parables and other humanistic components as essential. These help novice scientists realize that in essence science is not so much technical as architectural. There is also a secondary benefit, especially for the freshman course. The class is large enough that even freshmen not enrolled in it, many of them determined to avoid science, usually have a roommate or a friend across the hall in "Chem Zen." To my surprise, some of the unenrolled students began coming to lectures and to my office hours. Their interest had been aroused by liberal science material that prompted dormitory discussions.

The many efforts now underway to improve science education and public understanding of science acknowledge how immense and recalcitrant is the task. A recent book by Morris Shamos, *The Myth of Scientific Literacy*,[4] gives a comprehensive review and assessment. He concludes that only 5 to 10 percent of Americans can be considered literate in science or mathematics. I remain optimistic, in part because of my experience with "Chem Zen," and will suggest some ways in which more scientists and science teachers can readily contribute. First, however, I quote appeals from two revered scientific sages. In *On Understanding Science*,[1] soon after the advent of the atomic bomb, Conant wrote:

"Is it not because we have failed to assimilate science into our culture that so many feel spiritually lost in the modern world? So it seems to me. Once...assimilated, [science] is no longer alien...[and] becomes an element of strength...[Then science] will be fused into the age-old problem of understanding man and his works.

We need a widespread understanding of science in this country...
[It will bring us] one step nearer...[to] a coherent culture suitable for our American democracy in this new age of machines and experts."

Another fervent plea was made 40 years ago by Isidor Rabi in a lecture:

"To my mind, the value of science or [the humanities] lies not in the subject matter alone, or even in greater part. It lies chiefly in the spirit and living tradition in which these [different] disciplines are pursued...Our problem is to blend these two traditions...The greatest difficulty which stands in the way [is] communication. The nonscientist cannot listen to the scientist with pleasure and understanding.

Only by the fusion of science and the humanities can we hope to reach the wisdom appropriate to our day and generation. The scientists must learn to teach science in the spirit of wisdom, and in the light of the history of human thought and human effort, rather than as the geography of a universe uninhabited by mankind."

This and much more is given in a fine biography, *Rabi, Scientist and Citizen*,[5] by John Rigden.

My suggestions stem from the root problem: our society does not regard science and mathematics as part of our general culture, but rather as the province of priestlike experts. This attitude is indulged by teaching science as wholly separate from other subjects, too often in a ritualistic way. We need not acquiesce in this disjunction. Much more can be done to foster the assimilation and melding urged by Conant and Rabi, if a sizable fraction of the 5–10% of our science literate citizens becomes evangelical. In brief outline, here is an agenda:

(1) In developing liberal science themes, teachers can equip and encourage science students to be ambassadors to nonscientists, as happened inadvertently with "Chem Zen." The enthusiasm

of students can promote understanding both among their fellows and their families.

(2) Teachers offering either "hard science" or "physics for poets" courses should seek out faculty colleagues willing to include appropriate scientific parables in nonscience courses. A favorite example: the very popular Harvard core course on East Asian Civilizations, nicknamed "Rice Paddies" by students, could tell why these cultures rely on rice rather than wheat; it is a striking science story with links to much else.

(3) University professors should encourage graduate students to include two unorthodox chapters in Ph.D. theses: one reporting an initiative or experiment in teaching; the other describing the research in language comprehensible to nonscientists.

(4) Members of professional societies can promote and join programs for public understanding of science. Among others, the American Association for Advancement of Science, Chemical Society, Physical Society, and the National Academy of Science now all have such programs (with web pages).

(5) Scientists should ensure, by request or donation, that local media, libraries and other public sites receive suitable materials. In particular, the weekly *Science News* provides excellent accounts, nontechnical and sprightly, of current discoveries and issues. This is now augmented by a web site (http://www.sciencenews.org).

(6) Scientists can readily find opportunities to present demonstrations, talks, films or videos at schools and public forums. Science museums especially welcome help in presenting entertaining and instructive displays and programs.

(7) With humanist colleagues, scientists can form reading groups to explore books of mutual interest. Excellent science writing with a humanistic focus is now abundant. My reference list for this article has examples, mostly of recent vintage. These include history of science,[6,7,8,9] insightful exposition,[10,11,12,13] lyrical essays,[14,15,16] and even novels incorporating genuine science.[17]

(8) Whether at academic institutions or not, scientists can serve as organizers or judges for high school science fairs. These come in

local, state, and regional varieties, from which finalists are selected for two major events: the Westinghouse Science Talent Search and the Intel International Science and Engineering Fair, both administered by Science Service (youth@scisvc.org). The IISEF has now grown to over 1000 contestants from about 50 countries, requires about 1000 judges, and distributes dozens of scholarship awards. Anyone in need of optimism about the future should witness the student projects exhibited at such events.

(9) An alert scientist will find many informal openings for liberal science, without imposing on listeners. For instance, as the jumbo jet rumbles along the runway, I often try a gambit like: "This plane weighs 350 tons—I'm always amazed that it can take off, even though I'm a scientist!" Usually my neighbor then asks how flight *is* possible. That leads to the water pump story and beyond.

(10) A most important public responsibility, for individual scientists as well as science organizations, is to blow the whistle on fraud or pseudoscience and oppose attacks on valid science. Such things are rife today; many are examined in *The Demon-Haunted World* by Carl Sagan,[18] and in a recent symposium volume, *The Flight from Science and Reason*.[19] More scientists will have to respond, with efforts ranging from letters to newspapers to political contests against creationist candidates for local school boards...or perhaps higher offices!

This list of labors does not require herculean exertions; for the most part, it calls for happy rather than heroic actions. The chief point is that, according to taste and temperament, many scientists, teachers, and science fans can contribute in worthy ways. We may fall short of the lofty aims voiced by Conant and Rabi. However, we can share more fully "what's left" by evangelical efforts for public understanding of science as a human enterprise.

Notes and References

1. Conant, J.B. *On Understanding Science.* New Haven: Yale University Press, 1947.

2. Gordon, J.E. *The New Science of Strong Materials, or Why You Don't Fall Through the Floor*. Princeton University Press, 1968.

3. Flatow, Ira, *Rainbows, Curve Balls, and Other Wonders of the Natural World Explained*. New York: Harper & Row, 1988.

4. Shamos, Morris, *The Myth of Scientific Literacy*. New Brunswick: Rutgers University Press, 1995.

5. Rigden, S. John, *Rabi, Scientist and Citizen*. New York: Basic Books, 1987.

6. Cohen, I.B. *Science and the Founding Fathers*. New York: W.W. Norton, 1995.

7. Holton, Gerald, *The Advancement of Science, and Its Burdens*. New York: Cambridge University Press, 1986.

8. Holton, Gerald, *Einstein, History, and Other Passions*. New York: Addison-Wesley, 1996.

9. Sobel, Dava, *Longitude*. New York: Walker, 1995.

10. Chaisson, Eric, *Cosmic Dawn*. New York: W.W. Norton, 1988.

11. Diamond, Jared, *The Third Chimpanzee*. New York: HarperCollins, 1992.

12. Morrison, Philip and Phylis, *The Ring of Truth*. New York: Random House, 1987.

13. Trefil, S. James, *Meditations at Sunset*. New York: Collier Books, 1987.

14. Bronowski, Jacob, *Science and Human Values*. (Rev. Ed.) New York: Harper & Row, 1965.

15. Cromer, Alan, *Uncommon Sense*. New York: Oxford University Press, 1993.

16. Hoffmann, Roald, *The Same and Not the Same*. New York: Columbia University Press, 1995.

17. Djerrasi, Carl, *The Bourbaki Gambit*. Athens, GA: University of Georgia Press, 1994.

18. Sagan, Carl, *The Demon-Haunted World; Science as a Candle in the Dark*. New York: Random House, 1995.

19. Gross, R. Paul, Levitt, Norman, Lewis, W. Martin, Eds. *The Flight from Science and Reason*. Vol. 775, New York Academy of Sciences, 1996.

8

Science Education and
The Crisis of Gullibility

Andrew Fraknoi

"The scientific education (as distinguished from the purely technical training) of our people has been neglected, and the result for all to see is the almost unbelievable success obtained by quacks, faddists, and cultists of every kind. That success is possible only because of the poverty of the scientific spirit, the lack of resistance to irrationality. It is urgent to build up that resistance ... In fact, there are a good many people who believe themselves to be scientifically minded, because they can make a show of scientific knowledge (often the latest and the most controversial), yet have no definite idea of scientific method... it is very necessary that the [average person] should understand as clearly as possible the purpose and methods of science. This is the business of our schools, not simply of the colleges but of all the schools from the kindergarten up."
George Sarton in the introduction to Bernard Jaffe's *Men of Science in America* (1944, Simon & Schuster)

Or, put another way, *"There's a sucker born every minute."*
Phineas T. Barnum (1810–1891)

I would like to suggest that one of the most detrimental (and least discussed) effects of the crisis in science education in the world today is that we are creating a population increasingly unable to think *skeptically* about a wide range of issues. This is happening at just a time when the issues confronting society are becoming more and more complex, and thus the opportunities for deceptive claims and outright fraud are greater than ever.

Now I want to be sure that we don't confuse cynical and skeptical thinking. Any of us who teach (at least in the U.S.) can attest to the fact that the generations now in college are perhaps the most cynical in history. They have few heroes (and the ones they have are generally entertainment or sports figures because that's what the media dish out to them) and they seem to believe in little except the importance of personal pleasure and wealth. They sense that many political leaders wind up betraying or at least subverting the public trust for personal gain or to obtain funding for reelection and thus frequently express very little faith in the political system. Many have searched for role models and mentors, but few have found them in the larger world to which they are daily exposed. No wonder that it's a rare student who comes to us with his or her idealism intact.

But behind their cynicism, students (like so many adults) hunger for something larger than themselves and their daily cares, for systems that will imbue their lives with meaning. This hunger, frequently unsatisfied by some of the traditional religious and philosophical perspectives (at least in the U.S.), has led to the current rise of evangelical and fundamentalist religious groups, cults of all kinds, new age organizations by the dozens, and charlatans ready to make a nice profit from the confusion and rootlessness so many people feel.

At such times, it would seem to be the clear responsibility of leaders in all fields, and for teachers at all levels, to help educate young people on how to make difficult decisions on such issues, and to equip them with the intellectual tools needed to investigate and choose among the many conflicting claims with which they will be confronted. Yet, my own experience is (and surveys confirm) that, after 12 to 16 years of education in the U.S. our students are instead more gullible than ever. (The same, unfortunately, can also be said about your average television news reporter; more on that in a minute.) By gullible, I mean ready to believe the most amazing assortment of non-sense, pseudo-science, rumor, and innuendo, and lacking, in many cases, the tools for distinguishing between these ideas and the real world.

Defining the Problem

What sorts of ideas do I mean? Notions like astrology, the ancient view that the position of celestial objects at the moment we were born influences (or even determines) our love life, our choice of career, and our general destiny. One out of four adult Americans believes this is true, and 22% aren't sure whether it is or it isn't, according to a 1990 Gallup Poll.[1] Twenty-seven percent believe that "extraterrestrial beings have visited the Earth," and 32 % are not sure. Almost 50% answered yes to the (admittedly vague) question, "In your opinion, are UFO's something real?"

These days, a psychiatrist associated with one of the Harvard University teaching hospitals, John Mack, is getting an enormous amount of publicity (and, not incidentally, making quite a bit of money) by claiming that some of his patients were abducted by UFO's and subjected to sexual and other physical experiments while aboard. Then there is the idea, popularized by no less a scientific authority than Shirley MacLaine, that some people can "channel" spirit-beings from the past or another plane of existence. Or the now declining, but once very widespread notion, stirred by a convicted Swiss embezzler named Erich von Daniken, that ancient astronauts from another world had to help us start civilization, because our own ancestors were too stupid to do it themselves.

It seems to me that even a few decades ago, the institutions of our society had better control over the dissemination of such "fiction science". Certainly, the mainstream publishers, television news organizations, and the daily press had publishers, editors, and reporters who would have been disgusted and ashamed to be associated with these sort of ideas. While there is a long tradition of such media as *The National Enquirer*, for the most part they were confined to a kind of "backwater" of journalism and had a much smaller influence on the average student growing up in America. Today, tabloid journalism seems to be the rule rather than the exception; the main television networks vie with each other as to who can get the most sensationalistic and absurd UFO or psychic power stories on their news magazine shows. And the same publishers who issue science books by Stephen Hawking or serious literature by John Fowles fall all over themselves to publish pseudo-scientific drivel.

Bombarded by a constant diet of pseudo-science (without hearing much about the other, skeptical, side of many of these issues), our students grow up believing that "if so many people believe in it, and talk about it, and so much of it seems so credible on television, well, there must be something to it."

Does Anyone Important Believe this Stuff?

At a meeting like ours, you may laugh at the notion of anyone in any position of power taking this kind of belief seriously. But let me give a few examples. When funding for NASA's microwave search for signals from possible extraterrestrial civilizations was being debated in Congress, at least one of the people's representatives said on the record that there was no need for such a search because UFO records revealed that aliens were already here and could thus be contacted at no expense to the government. In India this past week (September, 1995), huge numbers of people stopped the daily course of business, and parts of cities had to shut down when the story began to spread that certain statues of Hindu gods were drinking spoonfuls of milk held to their mouths.

For much of the 1980's, while humanity sent a spacecraft through the coma of Halley's Comet, while computing power that a decade before required a room full of computers became available for any desktop, while medical diagnostic tools permitted unprecedented sophistication in ferreting out life-threatening diseases, a wily astrologer had the First Lady of the U.S. believing that her charts could lead to better decisions about the President's movements and meetings than the work of the entire executive branch of the government.

The White House chief of staff, Donald Regan, wrote later: "Virtually every major move and decision the Reagan's made during my time as White House chief of staff was cleared in advance with a woman in San Francisco who drew up horoscopes to make certain that the planets were in favorable alignment for the enterprise... Although I had never met this seer,... she had become such a factor in my work, and in the highest affairs of the nation, that at one point I kept a color coded calendar on my desk as an aid to remembering when it was propitious to move the president of the United States from one place to another, to schedule him to speak in public, or commence negotiations with a foreign power."[2]

If a large number of our citizens (and even our leaders) believes that our lives are governed by and decisions can be made by magic and superstition, will they feel the same urgency we do about the need for greater understanding of science and technology to solve the difficult problems of our age?

Some Solutions

It seems to me that any attack on the problem of increasing gullibility must occur on two separate fronts: the schools and the media. As has been discussed in many places, one of the most troubling aspects of the crisis of scientific literacy in our schools is that so many teachers continue to portray science as a collection of facts, rather than a technique for coming to understand the world. So students may know a bit of science trivia when they graduate, but they miss the "big picture" of why science has been so effective and how science weeds out good ideas from bad using experimental tests. A receptive, but skeptical frame of mind is a key element in scientific thinking, and teaching about the scientific method is an excellent way to encourage skeptical thinking in many fields.

When I was a junior high school student, the single course that made the greatest impression on me was a social studies unit on deceptive advertising, in which our young and enthusiastic teacher explained with great gusto all the different ways that commercials and ads can lie, distort, understate, exaggerate, and massage the information that is conveyed to you. He helped us focus on the sorts of deeper questions one should ask and information one should gather before accepting claims that sound good on the surface. I only wish I could prescribe such a course to every student in America, especially before political campaign seasons.

But even if we could change the way science is taught in the schools (and many curriculum reform efforts that help students think are now under way), that would only be half (or less) of the battle. A typical student in the US spends about 900 hours a year in school and between 1200 and 1800 hours in front of a television set.[3] In 1993, the average US household had a television set on for about 8 hours per day.[4] And what does a youngster learn from commercial television—often the exact opposite of what we try to teach them in school:

- That the only thing that matters is the instant gratification of every urge ("just do it," as the popular slogan says);
- That what counts in America is money, celebrity, and fun;
- That anything can be true and can happen (and that anecdotal evidence is enough to prove any assertion.)

Think of how teachers are shown on most television programs on the commercial services: mostly as objects of derision. And how are scientists usually portrayed (when they are shown at all)? Mostly as evil villains, or naive agents of serious catastrophe that result from their experiments getting out of control. These "lessons" of television viewing are rarely lost on our students!

As a country, we need to recognize the enormous power that the media have, and pay more attention to how decisions are made in the media. By this I don't mean censorship, but rather undertaking a long-term educational program to make sure that those who have the responsibility for deciding what goes on the air (or what is printed) have the education and the informed background to make sensible judgments. One step would be to introduce and require excellent science overview courses developed specifically for students going into journalism. Another would be for everyone who is dissatisfied with science coverage on the media to make his or her voice heard locally or nationally.

The argument the media always make in response to such concerns is that they are merely giving the public what it wants. But this argument is specious. Someone like me, who began life in Communist Eastern Europe, never got to taste a mango or an avocado as a child. Thus I would never have known to ask for a mango or an avocado or that I wanted one. But after we came to America, I was introduced to a much wider range of foods: my palate was educated...and now I ask for mango and avocado regularly. Similarly, I would suggest that when the media tell us that people are much more interested in the Bermuda Triangle than the Great Attractor, they are merely confessing our joint failure at educating both the gatekeepers of the media and the public.

In focusing so strongly on the entertainment aspect of their mission over the educational aspect, the media lose sight of the fact that there is an alternative to pandering to the lowest taste. It is the far more difficult and long term job (and the almost forgotten pleasure) of elevating the tastes and desires of their audience to new levels of perception and understanding.

I should mention that there is a national organization which is making a creditable effort to help the media sort out science from fiction science. The group is called the Committee for the Scientific Investigation of Claims of the Paranormal (CSICOP), and consists of scientists, educators, magicians, philosophers, lawyers, and other skeptics whose interest is to get the rational, skeptical perspective about fiction science out to the media and the public. Through their superb magazine, *The Skeptical Inquirer*, their meetings and workshops, and their information releases to the media, CSICOP has managed to alert and educate a significant number of reporters to stop and consider what they are doing when they file a story involving pseudoscience. They also work to bring together such reporters with skeptical spokespeople before a story is written. (They can be contacted at: P.O. Box 703, Amherst, NY 14226; I urge everyone interested in science education to get to know them and support what they are doing.)

Putting more skeptical thinking into the curriculum and the media will not be an easy task. It will require fundamental changes in the way we teach science, from elementary school to graduate school. It will probably require a new approach to undergraduate instruction and to training for scientists, teachers, journalists, and business leaders. It may take voluntary codes of self-restraint for publishers and broadcasters, which raises very serious civil liberty and first amendment issues that will need to be debated carefully. But without some attention being paid to changing the way we teach our youngsters to reason and to judge, we may be ushering in an era (and it will not be the first in the history of the world) where the rational is enchained by the irrational, where reasoning gives way before authority or self-interest, and where the enterprise of science is seriously endangered. We allow such a world at our own peril.

Notes and References

1. Gallup, G. and F. Newport, "Belief in Paranormal Phenomena Among Adult Americans" in *Skeptical Inquirer*, Winter 1991, vol. 15: 137.
2. Regan, D. *For the Record*, Harcourt Brace Jovanovich, 1988.
3. Barber, B. "America Skips School" in *Harper's Magazine*, Nov. 1993, p. 39.
4. Wright, J., ed. *The Universal Almanac*, Andrews & McMeel, 1995.

9

False Experts, Valid Expertise
Ursula W. Goodenough

Why Study Science?

The argument for improving the quality of science education usually starts out stating that science education is essential if we are to train our youth for the jobs and opportunities of the twenty-first century. The image materializes of earnest, clear-eyed graduates, donning their lab coats or updating their software applications as they march into a future of success and financial well-being.

Nothing is wrong with this. Except that most people lack the talent, interest, and/or motivation to pursue such high-flying careers. Once most kids have taken a few science courses, they come to realize that other kids in the class are "better at it" than they are and the career rationale evaporates. If no other rationale is on offer, then there's no reason to continue the struggle of trying to remember what a hypothesis is.

A second rationale, then, is to argue that science is so fascinating that its mastery represents one of life's deep joys.

Nothing is wrong with this either except that it also doesn't work for most people. Most people, as near as I can tell, are interested in two things: their material well-being and their human relationships. The wonders of science just don't cut it. In fact, if anything, the worldview provided by science makes many people feel pretty uncomfortable. It talks of big bangs and suns that go out and genes that govern behavior. Even when not vocalized, there are many who wish that science would just go away.

So a third rationale is to argue that a knowledge of the scientific worldview is essential if the citizenry is to navigate the twenty-first century. One needs a core understanding of science to judge policy issues on the environment, to make appropriate decisions about health and nutrition, to comprehend DNA evidence in legal trials and advice provided by genetic counselors. Science, in this view, is a commodity, something of use, like being able to manage one's financial affairs (the homologous rationale for mathematical literacy).

Seeking Scientific Expertise

The key problem with this rationale is that it is hopelessly unrealistic. It simply isn't possible to make valid judgment calls on scientific matters with the incomplete understanding of science that most people have. I can speak of this first-hand from my very incomplete understanding of mathematics. When I hear an argument based on mathematics, I am usually incapable of evaluating it. I can recognize a number of its features—an integral sign here, three functional variables there—but I don't really get it. I need someone to walk me through it.

Whom do I ask to walk me through it? Well, a competent mathematician of course, like a friend in the math department at my university or the clever computer-whiz graduate student down the hall. And the same process occurs when, say, a physicist friend realizes that she doesn't understand the basis for DNA fingerprinting and calls me up. My understanding here is deep, and we talk until she is satisfied.

So if we professionals recognize such inadequacies and rely regularly on others to help us understand, how can we imagine that a highschool graduate, even after lots of good science courses, will be able to evaluate a novel scientific proposition? Yes it is crucial to have heard the basics, to know by rote the general way that integral signs and DNA fit into the scheme of things. Without this much background, any explanations are incomprehensible. But we all still need to ask the experts.

Asking the experts is not, however, a straightforward proposition. There are two key problems: the problem of the false expert, and the problem of the prejudices harbored by the seekers of expertise. Together these breed the opportunity for deception.

Deception and Prejudice

We all know how deception works. There are two players, the huckster and the dupe (in our examples we can assume them both to be generic males.) Each brings to the interaction a set of motivations and a set of understandings.

The huckster in this case poses as an expert on some scientific topic. He may in fact lack such expertise, or he may wish to promote an invalid scientific claim. His motivation is presumably to enhance his financial and/or social status, these being the motives for most deceptions. Thus the politician wants to convince the dupe that some land is being irrigated for the good of the local ecosystem, when in fact it is in the interests of local agriculture to do so. Or the marketer of herbal teas wants to convince the dupe that the teas will cure his arthritis, when in fact no such cure has been documented. The radon exterminator has a briefcase full of "scientific facts" on the harmful effects of radon. And so on.

The dupe is usually the more complex and interesting participant. The dupe brings to the interaction his endowment of scientific understanding which, as argued above, is probably incompletely encoded in a series of icons—integral signs and DNA—that don't cohere very well. The skillful huckster can and does work with this fragmented material, keeping his claims simple and using familiar phrases: "DNA, the carrier of genetic information" and "our precious heritage, the diversity of species". The idea is to come up with a deceptive spiel that plays to the fragmented understanding of the dupe and convinces him that he has understood (and hence agrees with) the argument.

We should pause to emphasize this important point. For something like science, where understanding requires cognition rather than intuition, the experience of understanding a proposition usually generates a tendency to agree with it. "I heard a man giving a speech about the dangers of radiation coming out of power lines and you know, he made some good points that I could really follow. He explained magnetic fields a lot better than my high-school physics teacher ever did and I could understand how those fields might really be dangerous."

So the huckster can take advantage of the dupe's cognitive frame of reference. But fully as important is the huckster's understanding of, and hence manipulation of, the dupe's attitude towards science.

Although science is in fact nothing more than an account of the workings of Nature, it has acquired an additional status. Science is also something that one can have an opinion about. Thus science can be exalted, touted as the key to all problems, the engine of economic growth, the apotheosis of Western Civilization. Or it can be deconstructed, relativised as just another truth claim to take its place alongside the many other claims on offer. Science can be vilified, blamed for our excessive materialism and for the grave problems wrought by technology (the Unabomber critique). And finally, science can be feared, feared as the source of modern alienation, feared because it appears to demean our lives and challenge our centrality.

The dupe, then, will likely hold one or more of these opinions about science, which I believe we can fairly call prejudices about science. And the huckster, if he's any good at all, will manipulate these prejudices in developing his claims. Thus the dupe emerges from the huckster's speech with additional conviction: "And I always thought those power lines were dangerous anyhow. It's just one more thing that the utility monopolies are foisting on the public."

Humans are quite aware of the likelihood of being deceived. We understand that the radon guy and the herbal tea guy are trying to make a buck and that the politician is trying to get votes. We understand, that is, that these people are motivated to lie, and we believe that we have the resources, based on experience and gut feeling, to judge a person's sincerity ("Would you buy a used car from this man?"). But we lack the resources for judging the validity of most novel scientific arguments.

My sister-in-law speaks of this dilemma with real poignancy. She is an environmental activist, working to restore the Missouri River to its natural state. She reports that at numerous meetings and hearings, after she and her group have made their case, they are barraged with science-based arguments from those who would preserve the status quo. "When they do that we're helpless!" she moans. "We have no idea what they're talking about, and no way to argue against them."

Scientists as Hucksters

Particularly tricky are the cases in which scientific deceptions are perpetrated by scientists themselves. In other words, there are scientists out there who are hucksters. True, not most scientists. I am fully convinced that most scientists make every effort to be scrupulously honest in their exploration and explanation of the natural world. But it only takes a few bad apples to spoil a barrel.

Two examples can illustrate present-day manipulations of scientific validity and prejudice by scientists. The first involves the Space Station. The federal government is currently committed to spend some $90B to construct a Space Station. The one compelling reason for building the Station is that it will funnel monies into the flagging aerospace industry. Otherwise it has no rationale. A rationale was therefore devised. The Station, we were told, is necessary to explore "the new frontier" of microgravity, where bold scientific discoveries will be made on the effects of gravity on biological processes. When it was announced that research grants would be funded in this exciting new area, scientists materialized with projects galore. There is in fact little reason to believe that any of these projects are of value. Microgravity is, to be sure, an experimental variable, but not a very interesting or important one.

But once the ball got rolling it acquired a life of its own. Those supported to conduct microgravity research became the supporters of microgravity research.

And who are the dupes? Ultimately they are our political representatives who vote each year to fund the Station and the ancillary research grants. Many, of course, are simply voting to support the aerospace interests in their district. But their testimony reveals that they have ingested the prejudice of Science as the Engine of Progress. Hear, for example, members of Congress at a recent hearing.

"Space Station researchers will be able to use the low-gravity environment to expand our understanding of cell culture, which will revolutionize treatment for joint diseases and injuries". (Bentsen, D-TX).

"Scientists talk about how, in a gravity-free environment, they can make medicines that can help mankind… The Space Station is the frontier for America today. This is where the pioneers will go and this will bring us a lot of money". (Hansen, R-UT).

"What we have to do is put man in space in a space lab to do the kinds of wonderful experiments and scientific breakthroughs that come from that" (DeLay, R-TX). And so on.

A second example involves Dr. I.W. Lane, who flaunts his Ph.D. in Agricultural Biochemistry from Rutgers on the cover of his book called *Sharks Don't Get Cancer*. Lane uses this rationale to sell pulverized shark cartilage to terminal cancer patients, at thousands of dollars per victim, to be ingested orally or by enema. There are compelling scientific reasons to believe that this material cannot possibly be effective against cancer. Yet Lane makes hundreds of millions of dollars a year. Obviously a family devastated by a terminal-cancer diagnosis will be particularly vulnerable to deception. But Lane's approach is particularly effective. He emphasizes that his product (called BENEFIN) is "natural", in contrast to the "chemical" regimes of chemotherapy, thereby playing to the anti-science prejudices of his listeners. And he reinforces this by telling endless tales of how "the establishment"—pharmaceutical companies, the medical profession etc.—is trying to thwart his heroic efforts to bring this natural cure to those in need.

How to Proceed

To summarize, then, I argue first that the most plausible broad-based goal of science education is to render the public capable of evaluating science-based propositions. I then point out that such evaluations are rarely possible without the advice of experts, and that false experts can all too often use science-speak to deceptive ends. And finally, I note that persons usually develop emotional attitudes—prejudices—about science which can be manipulated and which cloud the evaluation process.

Therefore, while our schools should of course provide our younger children with opportunities for hands-on science exploration and our older children the vocabulary to recognize what DNA and integral signs are, we should also be talking in our classrooms about the importance of seeking valid scientific expertise and avoiding scientific prejudice.

Of key importance is that our children come to hold deep respect for academic institutions and an understanding that they represent the best source of scientific expertise. Americans are very confused about how to think about their universities. There is little recognition that to achieve the position of professor at a major university is quite as impressive as achieving the position of pitcher on a major-league baseball team. In both cases, the competition is fierce, the training is strenuous, the imperative to keep in top form is relentless. Our populace must come to understand that when they are offered scientific expertise from the academy, they are listening to some of the most distinguished and knowledgeable persons that our society has produced.

Having said this, of course, it is imperative that academic institutions earn this kind of respect. Specifically, scientists must come to look upon the provision of scientific expertise in the same way that physicians look upon the provision of medical care. That is, scientists carry the obligation to monitor their professions and blow the whistle on false experts and exploiters. Unfortunately, this is a responsibility that scientists are shamefully loath to take on. We roll our eyes at microgravity research or claims about shark cartilage, but all too many of us then look the other way. Some of this behavior is generated by laziness or busyness, but much stems from a robust and growing ethos in the scientific community which states that I won't criticize your science (and thereby jeopardize your funding) if you don't criticize mine.

We do this at our peril. We are the only ones, in the end, who can distinguish scientific hucksterism from scientific truth, and if our expertise is to be trusted, we must take the lead in driving the hucksters from our society.

But once we have our house in order and have earned the public's trust, it will still be difficult to offer valid expertise so long as the public harbors prejudices about science. The deconstruction of prejudice is obviously an across-the-board challenge for our educational institutions. But there seems to be far too little concern about exposing the prejudices that exist about science. The message provided by the schools can be very straightforward. Science is the best account we have of the natural world. Period. If we could teach our children that this is the case, and teach them that they can trust valid scientists, and valid scientists alone, to help them understand the physical dimensions of reality, a great deal would happen much more wisely.

10

Confronting Complexity:
A New Meaning to World Literacy
David Chen

Introduction

Knowledge has become the main element in a society heading into the twenty-first century. Accordingly, the present era has been variously termed the Information Society,[1] the On-line Society,[2] the Global Village,[3] the City of Bits,[4] and perhaps most fitting of all, the Knowledge Society.[5] It was Daniel Bell of Harvard who foresaw and characterized the advent of the Knowledge Society.

The Knowledge Society is one in which theoretical, abstract information (systems of symbols) becomes the main factor in the economy and most workers are employed in high-tech market sectors. Mitchell[4] and Negraponti[6] extended the description of the significance of this change to culture and human life-styles, partnership in democratic life, multidimensional virtual communities, and human interaction which is not dependent on time and place, and so on.

It is possible to discern an additional dimension of human existence in the Knowledge Society, one which Sir Karl Popper[7] called "World 3." World 3 represents exosomatic knowledge; that is, human information that lies outside of the brain. Its digital representation is referred to in current terminology as cyberspace. Today, in addition to the classic infrastructure of print and paper, a great deal of public information exists within a computerized technological infrastructure,[8] and it is available to anyone with access to a digital environment. A new era in human civilization has begun.

This new social, cultural and economic experience is exciting; it ignites the imagination and appears to offer a wealth of new opportunities to humankind. There is no doubt that the Knowledge Society presents new challenges to education, which until now has dealt mainly with print-encoded information.

It seems that full participation in the twenty-first century in the Knowledge Society means:

- free access to information for everyone
- the right to learn at any time and any place
- expanded ability to choose in learning,
- recognition of the cultural pluralism of knowledge,
- the ability to act on the basis of knowledge.

Is contemporary education prepared to meet the new demands of the knowledge society?

To cope with these challenges, long-term changes in educational goals are needed. A major portion of these changes are manifested in new relationships between personal and public-knowledge.[8]

The main aim of education was and remains the imparting of reading and writing skills, or the classic three r's: reading, 'riting and 'rithmetic. Literacy has always been considered to be the key that opens the door to culture, but as we draw nearer to the year 2000, it takes on a different, broader meaning. The three goals of this expanded literacy are:

- To know (perception of the world)
- To understand (ability to decipher the complex world)
- To act (to change behavior on the basis of understanding).

What is the significance of these goals to the education system? What must we do to achieve them? What difficulties lie along the way, and how might we overcome them?

The optimistic forecast of individuals and communities of future generations joining in a grand partnership in democratic life, enjoying economic prosperity and contributing to a rich culture, is confronted by a complex and grave universal reality. The education system all over the world is in crisis.

This universal crisis stems from the inability of educational systems to prepare for long-term social changes. The gap between school and life continues to widen, as the break with traditional "book learning" takes on a more radical significance in the cyberspace culture. Today, a considerable portion of children being born all over the world will have only a minimal chance of taking part in the Knowledge Society. Great responsibility for the next generation lies in the hands of today's leadership. These are the issues we will be dealing with in the present article.

Literacy in the Knowledge Society

A. The social problem

When the physicist and philosopher Charles Percy Snow [9] spoke of the "culture gap," he meant the gap between the humanistic and scientific communities. Although this gap has today taken on totally different, broader dimensions, a look at the global picture shows that, according to data gathered by the United Nations Educational, Scientific and Cultural Organization (UNESCO),[10] there are some 900 million people over the age of 15 who are totally illiterate, which constitutes 32.8% of that population group (3.25 billion). The inability to read and write relegates these people to an extremely low level of participation in civilization, beginning with personal income, and also including participation in democratic processes, enjoyment of a high standard of living, and contribution to culture. Moreover, the global distribution of illiteracy is uneven: the illiterate population of developing countries (Africa, Arab countries, parts of Asia) numbers approximately 873 million compared to only 31.5 million in developed countries (North America, Europe, parts of Asia/Oceania). These data may be further clarified by mapping out the rate of participation in educational systems worldwide. Based on the above, the world may be divided into civilizations rich in knowledge (inforich) and those poor in knowledge (infopoor). Such division creates even more significant cultural gaps than those pointed out by Snow.

Such a map describes the breakdown of the proportion of learning on the different continents. In large areas of Africa, Asia and South America, access to culture through education is extremely restricted. North America, Europe, Japan and Australia are inforich thanks to the high rate of learning which is prevalent there.

There is, however, more to the issue. The best of human civilization is anchored in language. While it is estimated that there are 6000 languages in the world, only 12 are used by more than 100 million people. Thus, information encoded in Chinese will be used by 1077 million people, in English by 593 million, in Hindustani by 412 million, in Spanish by 311 million, in Russian by 285 million, in Arabic by 206 million.

Written information is distributed proportionally among different cultures, whilst in contrast, computerized information exists mainly in English. The information highway operates in English, and technology has not fully resolved the problematic transition from one language to another. Therefore, cultures whose knowledge is encoded in a language other than English have only limited access to cyberspace; likewise, cyberspace has little access to information in other languages.

One could say that many of the problems of literacy in the Knowledge Society involve social, cultural and economic limitations. Accordingly, three levels of illiteracy may be defined:

• Basic illiteracy—deficiency in the ability to read and write in the mother tongue; applies to a significant portion of humanity.
• Secondary illiteracy—lack of access of information encoded in a language other than the mother tongue, mainly in English; isolates much of our collective information in cultural islands.

- Tertiary illiteracy—lack of access to information because of technological limitations. Knowledge that depends on a technological infrastructure such as computers and telecommunications which is not available to those who lack either the necessary resources to set up such an infrastructure (in developing countries or communities with few resources) and/or the necessary expertise for using information technologies.

Up to this point, the literacy problem has been outlined at the social, economic and cultural level. However, those who have already overcome the basic literacy obstacle encounter the educational problem, which is defined below.

B. The educational problem

It seems that the public debate on the "educational crisis" is, in effect, being conducted in those societies that boast a very high level of basic literacy—that is, the United States and Europe. The significance of this is straightforward: it may not be assumed that literacy at the basic level automatically ensures access to and acquisition of knowledge. Apparently, even in cultures in which the entire population shares reading and writing skills, serious education and learning problems are rampant. Despite all efforts to date, a major proportion of students fall short of the necessary achievements in the different areas of knowledge. The universality of this problem has been shown by comparative studies conducted by the IEA.[11] Low achievement has thus become the focus of the "educational crisis" in the West.

To cope with the definition of the educational crisis, the concept of literacy has been expanded from knowledge of reading and writing to mastery of basic concepts and skills in different disciplines: scientific literacy, technological literacy, humanistic literacy, civic literacy, and so on. There is today a movement to set standards of disciplinary knowledge in order to ensure that students reach an educational level required for living in the knowledge society of the twenty-first century. However, standards relate only to the symptoms of the crisis; its roots are partly social, partly cognitive.

Rather than trying to characterize the basic concepts and skills required for each knowledge domain, I would like to adopt an interdisciplinary approach in order to define the essence of literacy in the twenty-first century, and to understand the roots of the crisis. Without such a picture, it will be hard to formulate a strategy to achieve literacy.

There are four main topics on the educational agenda of the citizen of the twenty-first century:[12]

i) Understanding the relationships between people and their natural environment

The accelerated advance and development of civilization has upset the complex balance of relationships between people, animal groups and the planet on which we live. It seems that there are limits to growth, and we must adopt an approach of sustained development to cope with environmental problems. Environmental education is vital to fostering literacy, which requires becoming well-acquainted with the laws of natural science.

ii) Understanding the relationships between people and society

Human society has always been multifaceted. The relationships between the individual, the community and the rest of society are set by different arrangements. Democratic administrations strive for a fair balance between private and public rights, but their political, economic and organizational processes are dynamic and demand constant involvement by the individual. Civics education is a cornerstone of every democratic country.

iii) Understanding the relationships between people and technology

Technology has been developed to such an extent that it has become a central factor in our environment. The artificial has replaced the natural.[13] Technology makes human existence possible; it plays a role in food production, defense, and longevity; it furnishes material resources and energy, but at the same time, it endangers human existence. Only an understanding of the essence of this new technology, its advantages and limitations, will enable humankind to use it as a means and not an end.[14]

iv) Understanding the relationships between people and culture

Culture is a pluralistic product of human creation. It may be local or universal; it is rich and varied and displays many ambiguities; and the transition from tribalism to the global village has shown historic, and sometimes apocalyptic, dimensions. Instilling an understanding of human culture's many faces, together with the ability to actively take part in and contribute to it, is a time-honored task of the education system.

These four topics constitute the main context of literacy education for all those who are learning. Their common denominator is complexity. Environment, technology, society and culture are all, according to Forrester's definition,[15] complex, multivariable, dynamic, nonlinear systems that sustain many reciprocal connections. Forrester maintained that these systems are difficult to define conceptually and that most people are hard put to comprehend them.

In order to understand the different dimensions of complexity and to properly evaluate the cognitive demands which they make on the learner, I shall briefly elaborate on them below:

1. *Scale*. Science perceives a multidimensional world on a metric scale of 10^{42} orders of magnitude—from 1 billion light-years to 0.1 Fermi—as compared to the 10^8 orders of magnitude that are grasped by human senses. The immediate world perceived by the senses, is thus much more limited than that perceived by science. Thus, to conceptualize the world, people must be able to perceive a very wide scale that depends on very

large numbers and to make abstractions that can be processed via the intellect, not the senses.[16]

2. *Hierarchies.* The world is composed of systems and subsystems which are structurally and functionally interrelated. The structure of systems and subordinations create the phenomenon of hierarchy. Understanding of these structures and arrangements requires systemic thinking.[17]

3. *Redundancy.* Various components of the world are duplicated many times over. This is true of language, the animal kingdom, the physical environment and human society. The ability to identify such structures allows people to classify the world and confer meaning on it. Redundancy facilitates the processing and storage of information through the application of general laws.

4. *Near decomposability.* Strong interactions between world subsystems are indistinguishable, whereas weak interactions enable pattern recognition.[18]

5. *Change.* Changes occur in time and space. Movement, behavior, transitions and transformations are dynamic phenomena which science (and measurement) endeavor to define and characterize. Piaget[19] discovered that learning change is one of the most difficult cognitive tasks.

The ability to understand complex phenomena is the foundation to building a modern perception of the world; it is, therefore, a major goal of education. Clearly, new cognitive demands will have different qualities from those that have served traditional education.

There is a significant difference between the development of science as a collective effort and the development of scientific concepts in the individual child.

The main strategy adopted by science for deciphering the world was inspired by Descartes's analytic approach,[20] namely, the complex world must be broken down into small, measurable, comprehensible units. It has proved quite successful, and the work of the scientific community over hundreds of years has shown some magnificent results, especially in the twentieth century. In the educational system as well, this strategy has been adopted as a fundamental principle for the teaching and learning of science. Students are taught about a world that contains very few variables that interact linearly in closed systems. However, we are asking that they learn within a short time what took the scientific community hundreds of years to achieve.

There is also the curricular aspect. Traditional curricula divides subject matter into disciplines, with each taught separately from the others. Modern learning requires an expansion into interdisciplinary studies, both in the curriculum and in the organization of the school itself and its methods of learning. It is no wonder that most students find it difficult to conceptualize a world picture formed by nature, society, culture and technology.

Recent studies[21] have indicated that in countries with a high literacy rate, scientific literacy is reached by only approximately 7% of the population. Given the scope of this educational failure, we are forced to pose yet again the cardinal questions regarding the functioning of the educational system.

The problem of imparting expanded literacy may be summarized as follows:

- The basis of scientific and social education is complexity, not simplicity.
- The study of complex phenomena places a heavy cognitive burden on students.
- To develop literacy intended for a "perception of the world, new teaching strategies must be formulated.

Courses of Action

The development of literacy in its broadest sense—a personal information infrastructure that provides a new "perception of the world"—is a fundamental condition for the participation of future generations in culture, society and economics. To guarantee

this kind of literacy, we need to operate along two complementary
planes: the social and the educational.

The worldwide cultural gap is becoming increasingly wider.
Millions of people lack access to information. Even according to
UNESCO's narrower definition, 30% of the world population is
illiterate. There is no doubt that the great demands of the modern
world are changing the meaning of illiteracy and thereby widening
its scope.

The international educational effort is spreading knowledge of
reading and writing. But what of the other types of illiteracy that
stem from little or no access to information? The bulk of this new
information is in English. Since English is the mother-tongue of
only 600 million people, this means that the remaining 4 billion
or more are information deprived. Some countries solve this
problem by teaching English as a second language; others confine
themselves to their own language and culture.

Technological solutions such as machine translation, even if
successful, are not the definitive answer because of the cultural
aspect, which forms an integral part of the function of the mother-
tongue.

The problem of access to digital information becomes even more
serious in an era in which much information of scientific and economic import is computerized. Economic gaps prevent developing
countries from establishing communication infrastructures, and
low per capita income prevents individuals and schools from
acquiring computers and communications equipment. This, of
course, creates a vicious cycle: the economic-cultural-social gap
leads to more and more illiteracy, whilst growing illiteracy further
widens the economic-cultural-social gap. Solutions lie on the
political-economic level.

The main components of educational strategy in the social arena
are:

- a policy that gives high priority to education;
- study of English as a second language;

- economic development, changes in the allocation of resources and the development of information technology infrastructures.

There are no easy answers to the problem of illiteracy at the social level and I am not suggesting that I can provide any. However, even if UNESCO and its member countries succeed in conquering basic illiteracy, and even if developing countries provide their people with access to information, we will still have to cope with problems that arise at the educational level.

Complexity has been characterized as an issue common to most areas that must be learned to achieve expanded literacy and a modern perception of the world. In the accepted educational tradition, curricula, teaching and learning all emphasize simplicity. The major failure lies in the study/teaching of complex systems. Whether we accept Tversky's[22] and Forrester's approach[15] that the problem of learning complex phenomena is innate and that the senses and brain are not programmed to process complex problems, or that of psychologists who claim that complexity causes an intolerable cognitive burden, we must turn our attention to the search for learning and teaching methods suited to complex systems.

Ultimately, it was the human brain that enabled consciousness to expand beyond elementary particles right up to the cosmos, and scientific enquiry now enables humans to understand complex changes in physical, biological and social systems. We believe the problem lies not in the possibility or impossibility of learning and understanding complex systems in a physical, biological or social context, but rather in how to teach this kind of system to the general student population.

One obvious course of action is to foster system thinking as part of the expanded curriculum. The Science in Technological Society program, developed in Israel, has been conducting interesting educational experiments emphasizing interdisciplinary fields of science, technology, and society.

Another method arose in the wake of Forrester's claim[2] that complex and dynamic phenomena can be studied only in integrated human-computer systems. The most outstanding example of this approach is the Stella simulation system. This learning environment was used for the study of epidemiologic phenomena, ecological problems, social planning and so on, and has been employed by several research and development groups. Resnick's Star Logo environment[23] teaches dynamic, nonlinear processes such as ant behavior or transport systems. Milenski,[24] working along similar lines, used gas systems to examine how people learn thermo-dynamics. However, in none of these experiments was a comprehensive, unifying perception developed which related to the problem of learning complex phenomena. Research into systems thinking is still in the preliminary stages. Most publications deal actively with topics taken from systems theory, but not with the problems of perception and cognition involved in systems learning in general and complex systems learning in particular. Thus, despite the great interest in "complexity" shown by the scientific community,[25] almost no research has been done on the subject of the perception and learning of complex systems.

Regardless of whether the search will be in the direction of conventional or integrated human-computer learning environments, the principle task of education in the next century will be to overcome the problems inherent in learning complex and interdisciplinary phenomena.

Conclusion

The main goal of education in the twenty-first century is expanded literacy for all, which has become the prerequisite for the full participation of coming generations in cultural, social and economic life.

Literacy in its new sense requires the transition to developing the ability to decipher texts and studying the four main contexts into which human beings play the central role: the natural environment, the social environment, technology, and culture. Our awareness of these systems and the reciprocal relationships within them requires the ability to study and understand the phenomenon of complexity.

The strategy used to achieve expanded literacy for all must include two different yet complementary components:

On the social level: There is a need to overcome basic illiteracy, which remains very prevalent, predominantly in developing countries. Secondary illiteracy, the source of which are cultural and linguistic islands, must also must be conquered, mainly by fostering the use of English as a second language in non-anglophone cultures. Technological infrastructures must be built to allow widespread access to digital information, at least within public educational systems.

On the educational level: Efforts must be directed at developing innovative curricula and learning environments which place an emphasis on the ability to decipher complex systems within a context of an interdisciplinary perception of the world, anchored in science, society, technology and culture. This trend demands an investment in research and development, international cooperation and, especially, mobilization of the intellectual resources which will make it possible to develop a new educational agenda.

Notes and References

1. Daniel Bell, "The social framework of the information society." In Michael L. Dertouzos, Joel Mosses (Eds.), *The computer age*, 1980.
2. Tom Forrester (Ed.). *The microelectonics revolution.* Cambridge, MA: MIT Press, 1981.
3. Marshall McCluhan, *Understanding media: The Extension of Man.* Signet Books, 1964.
4. William J. Mitchell, *City of Bits.* Cambridge, MA: MIT Press, 1995.
5. Peter Drucker, *The Rise of the Knowledge Society: Dialogue*, 1994.
6. Negraponti, N. *The Digital Man.* Cambridge, MA: MIT Press, 1995.
7. Karl R. Popper and J.C. Eccles, *The Self and Its Brain.* Springer-Verlag, 1977.
8. David Chen, "An epistemic analysis of the interaction between knowledge, education and technology." In Ed Banner (Ed.), *Sociomedia*, 1993.

9. Charles P. Snow, *The Two Cultures and A Second Look*. Cambridge University Press, 1959.

10. Fredrico Mayor, *World education*. MIT Press Report—UNESCO publishing, 1993.

11. International Evaluation Association (1984–1993). *International studies in educational achievement series*. Pergamon Press.

12. David Chen, "A design for a working model of a school of the future." In D. Chen (Ed.), *Education Towards the 21st Century*. Ramot Press, 1995.

14. Jack Ellul, *The Technological Society*. Vintage Books, 1964.

15. Jay W. Forrester, "Counterintuitive behavior of social systems." In collected papers of Jay W. Forrester. Wright Allen Press, 1975.

16. Philip Morrison, Phylis Morrison & The Office of Charles and Ray Eames, *Powers of Ten*. Scientific American Library, 1982.

17. David Chen and Walter Stroup, "General system theory: Toward a conceptual framework for science and technology education for all." *Journal of Science Education & Technology*, 2 (3) (1993).

18. William Bechtel and Robert C. Richardson, *Discovering Complexity*. Princeton University Press, 1993.

19. Jean Piaget, *The child's conception of the world*. Littlefield Adams, 1951.

20. Rene Descartes, *Discourse de la methode*, 1637.

21. Gabriel Garceles, "World literacy prospects in the turn of the century." *Comparative Education Review*, 3(1) (1990).

22. Tversky, A. and D. Kahaumon, "Belief in the law of small numbers." *Psychological Review*, 76 (1971): 105–110.

23. Mitchell Resnick, *Beyond the Centralized Mindset: Explorations in Massively Parallel Microworlds*. MIT Press, 1994.

24. Uri Wilenski, *Connected Mathematics—Building concrete relationships with mathematical knowledge*. Doctoral Dissertation, MIT, 1993.

25. Gregoire Nicolis and Ilya Prigogine, *Exploring Complexity*. W.H. Freeman and Co, 1989.

11

Sciences and the Future of Human Culture

Vilmos Csanyi

Tribal warfare in the Balkans and the global information superhighway are the two extremes of the overbred human population's behaviour: primitive impulses rooted in our evolutionary past on the one hand, high technology serving global communication on the other. The interactive system comprising both the biosphere and human society is in a highly unstable state, approaching a bifurcation point. Of the two possible paths, one may end in a moral and ecological catastrophe and the other may lead the human race into a path of a dynamic self-improvement. The prerequisite for the latter is the emergence of an altogether higher form of moral consciousness to control the critical processes of the global system for the sake of future generations.

Proposing a constructive course of action for the human race first demands clear understanding of the biological and cultural processes that have led to the current situation and an assessment of all the preconditions for the desired changes.

Factors from the Past Continuing to Exert an Influence
Evolutionary theory is one of the most successful theories of science. It provides the framework for understanding the most important biological phenomena: the emergence and interaction of species, the development of the biosphere and its self-sustaining operation. The concepts behind evolutionary theory also lend

themselves to interpreting the more general class of replication-based systems, such as cultural, technological and certain abstract systems.[1] Evolutionary theory is thus successful in explaining the emergence, culture and social institutions of mankind.[2,3]

Humankind's distinctive biological traits are the product of the approximately five million years separating his evolutionary path from that of the other anthropoids. Our human predecessors lived in small, highly social units, the group individuality of which set them apart from the other animals.[4] Any form of individuality accelerates the evolutionary process. In the early *Homo* units a characteristic new form of cooperation emerged based on a jointly developed action plan. The action plan is a mental construction, modelling expected events, defining the role of each individual in the forthcoming action, dividing the action into functional parts and taking various possible outcomes into account.[5] Overall this is a vastly more efficient form of cooperation than that which takes place within other social species. The action plans developed by human units are always peculiar to the group and this intra-group individuality provides the basis for selection between groups, thus accelerating cultural evolution.[6]

Mimetic Cultures

Devising an efficient action plan is mostly a communication exercise, but although there is an obvious increase in efficiency provided by language, the ability to speak is not a necessary precondition for the action-plan. Well rounded hypotheses exist which suggest the appearance of biological communication mechanisms, enabling the synchronization and planning of group actions, well before the advent of language. Mimetics are one such mechanism, comprising some 150–200 different possible messages and exceeding the number of messages animals are able to communicate by about a factor of an order of magnitude.[7] Another such mechanism peculiar to humans is the erapathetic ability which enables us to sense the motivational state of his mates. Susceptibility to hypnosis also belongs to this group, enabling control of one another by means of a close emotional bond and according to recent findings this is not a unidirectional communicational channel but a bilateral one.[8,9] Imitation, that is, the copying of the behaviour of conspecifics independent of rewards or a goal which very rarely occurs in its pure form in the animal kingdom, is yet another example of peculiarly human communication.[10] Out of these forms of emotional communication grows the making of rhythmic noises, chanting and primitive music, rhythmic movements, dances and rites. All these help to closely synchronize the emotional and mental states of individuals in the group, facilitating the emergence of a unified will and coherent execution. They fulfill the role of a primitive form of language, denoting and differentiating objects and concepts, as well as synchronizing actions and defining individual roles. Together they can be regarded as a culture, that Donald[11] called "mimetic culture".

These factors and others, including prolonged and thorough socialization and unconditional commitment to the group, together account for the remarkable evolutionary success of the active and efficient *Homo* groups.[12]

The Creating Language

As indicated above, social characteristics of the *Homo* species underwent a series of fundamental biological changes. The social creature that emerged was bound to his group until death and acquired its culture. Individuals were only parts of a "group-self" which synchronized their behaviour.[13] They were born into the protective bubble of a group culture within which they spent their entire lives, and from the closed perspective of which human outsiders, with their own group-specific customs and channels of communication, seemed as strange and incomprehensible as herds of wild animals. Out of this unconditional loyalty to one's own group, a biological predisposition towards xenophobia, i.e. the contempt of other groups, was born.[10]

The efficiency of the "group-self" as a whole depended on the efficiency of communication between its members. It is also clear that even pre-linguistic forms of communication were sufficient for the rudiments of conceptional thought, complementary cooperation and group action to arise. During the emotional synchronization necessary for united group actions, a network of emotional representations developed which made the socialization process of each individual complete and unequivocal.

More advanced evolutionary changes had to coincide with the appearance of the spoken word, making group cooperation even more efficient. The medium of advanced culture is the biological linguistic ability of the species. Through stories and myths it enables the past to be recalled, providing a mental and behavioural framework drawing on ancestral experiences. By linguistic means the past and the future can be created, natural phenomena can be decomposed to their constituents and new, imaginary structures can be constructed out of them. Language permits a particularly precise form of conceptual thinking to develop. Logic and analytic thinking can emerge, allowing the construction of a complex super model of the group and its environment, based on a large amount of evidence collected by the whole group.[14] This super-model can be regarded as a type of social group-mind, which is made up of the experiences and thought processes of individuals, but which has far greater autonomy than has any one of its members. The function of the individual is merely to experience, register and, using language, forward information to a higher level, where processing takes place. At this stage individual human personality is very undeveloped.

There is still much debate over whether language developed gradually over the course of several millions of years or whether it appeared suddenly, in the last 50,000 years. Either way, advanced cultures sprung up only after linguistic competence reached its present day level.

The thought processes of the primeval "group-mind" shifted into a higher gear and became more accurate when language appeared. A new level of evolution emerged: the evolution of thought or cultural evolution. Biological changes facilitating sociability also continued.[15]

The culture that develops in closed groups based on conceptual thinking is a very special kind of system. In the brains of individuals belonging to the group, beside the emotional network mentioned earlier, a huge semantic network develops which comprises all that the group can conceptualize: representations of actions, events, emotions and precepts, names, and abstract concepts. Almost every thought is the result of group processing: verbal communication. Compared to the lifespan of the individual, the evolution of such a culture is slow, since any new thoughts must originate in the brains of the small group and can only be added to the existing network if logical analysis demonstrates its meaningfulness, that is, if it is not found to contradict some other part of the system. This coherent conceptual framework together with the emotional network are exactly those which the individual adopts as cultural heritages during the process of socialization.

Any thoughts which might upset coherence are scrutinised by the group. This is why closed cultures are relatively conservative and develop slowly but provide a high degree of emotional stability to their members. Everyone has the same value-system, everyone has similar opinions and hundreds of years of precedents must exist for any course of action to be taken.[16]

Such a culture is efficient as long as it is stable, because only then is it possible to predict future events. This need for stability is the source of the phenomena best described by the term group transcendence. Language allows the group, the group-self to be named and thus begin a new, independent life. The existence of the group is not dependent on the existence of the individual; it was there before the individual was born and will continue when he dies. The existence of the group therefore, superceeding the existence of the individual, transposes itself into an imaginary virtual sphere and becomes an external reality for members of the group. The abstract concept of the group can acquire different properties; it is tied to emotions, it can be an actor in myths and a history of its origin can be created. The virtual world of myths can be populated by imaginary beings such as ghosts, demons and gods. Out of group transcendence comes the transcendence of the individual, creating the immortal soul in virtual space. In this way, group transcendence gives rise to the demand for and practice of spirituality.

Commitment and Negotiation

As those groups that had verbal cultures came to dominate the entire planet, new problems followed in the wake of their success. Some successful groups split, producing different groups with similar cultures, thus ending their initial isolation. Groups started interacting, fighting or negotiating for resources. The black and white behaviour patterns of unconditional commitment to one's group and xenophobia were supplemented by the bio-cultural abilities to negotiate and compromise.[4] The rules for making deals with outsiders are quite different from rules governing intra-group behavior. Total commitment, unconditional loyalty and susceptibility to indoctrination characterize the latter, whereas deceit, cunning, and exploitation for unilateral advantage are permissible in the former. The tendency to exploit other groups grows in proportion to the differences between them, until groups begin to appear to each other as contemptible aliens outside the sphere of moral regard, merely something to be eradicated. During the course of human evolution, the tendency towards antagonism on the one hand and negotiation on the other came to an equilibrium because the weapons of destruction were limited and the negotiating instinct is relatively strong. This positive equilibrium lays the foundation for the appearance of social structures greater than the original group, namely tribes, nations and states. Overlapping, nested and mixed groups also appeared. This posed a new challenge for evolution.

Pseudo-groups and the Autonomy of Modern Humans

In the time of closed group-culture the process of socialization was undisturbed and complete. Those born into any given group would never have to confront logical and emotional contradictions in their mental life, and the network of cultural conceptions seemed to be perfectly seamless, at least from the inside. Later, members of various groups were being forced to live in close proximity to each other, or indeed within each other's area, and the apparently seamless conceptual network of each group began to crack. Interacting with alien cultures gave rise to new questions, some of which could not be answered. These cross-cultural interactions not only enriched but endangered, by presenting contradictions and the possibility of unprecedented, formerly unimaginable actions. The crux of the new problem was that the individual would now have to make decisions and choices between good and bad, on his own. However, man was not biologically equipped for this, all decisions in the past having been made by the whole group. What is more, the systems for socialization and emotional stability suffered seriously as a result of this exposure. From the moment of birth onwards, children were assaulted by a variety of often contradictory messages. Unquestioning loyalty to the group could no longer be so absolutely inculcated. The natural groups into which we were born and lived in harmony, were superceded by pseudo-groups of the ad-hoc alliances, working groups and various religious, professional or other organizations operating independently of each other. These organizations are not the product of natural socialization processes but, due to our species' exceptionally strong gregariousness, they are still more or less functioning social units. Once we lost the clear, rational, unambiguous, emotional warmth of an all embracing culture, we found ourselves marooned in a merciless world of individual decisions and their sometimes cruel consequences. This is the source of the modern prevalence of neurosis.

The response to problems posed by the mixing of groups was the emergence of the concept of personal autonomy. According to this theory the fundamental unit of society is the individual, not the group; individuals must try to preserve their autonomy, and their loyalty to any group may not go so far as to threaten this. To which group he or she belongs, is a decision largely determined by his/her personal interests. This concept most characteristic of western societies makes no allowance for the existence of a super-structure above the individual, to which he must remain forever loyal and for which he may have to sacrifice his life. Instead, it advocates the ideal of society as a voluntary association of autonomous individuals acting in accordance with pre-arranged mutual agreements.

Upon closer examination of the concept of personal autonomy, we find that it is practically identical to the negotiating tactics of a single-member group. Population growth has diminished the range of action of natural groups to such an extent that they now contain only a single individual. This autonomous citizen constantly strives to secure the best possible solution for his group of one, in the competition between similar one-member groups.

Making a Global-mind: Global Culture or Global Cultural Ecology?

For an historically very short period, personal autonomy, western-style individualism and the negotiating systems of independent individuals have been successful. The population has reached, and now probably exceeds, the carrying capacity of the planet and this success has a high price. Irreversible changes of Earth's environment now threaten mankind and all its future generations. The moral and psychological process which converted the individual from a contributor to the wise social mind, to a single-member group, threatens catastrophe. Single-member units incapable of loyalty, are therefore incapable of sacrificing their own interests for the sake of the species or the planet, without radical readjustment of their behavioural patterns.

Perhaps the process is still reversible. Several possible alternatives exist, two of which are particularly promising. One is the creation of a global culture. Provided that young people are exposed to a socialization process which depicts the whole of humanity as a single group and develops their loyalty towards that group, a new global "self", an *ultra-mind* embracing mankind within a coherent culture, might be born. The gigantic task of science here is the integration of global culture, by means of eradicating logical and emotional contradictions. The transcendence of such a global culture is the goal of that cosmic vision which sees man as an active creator of the Universe. This scenario relies on the cultural differences between groups gradually diminishing, local cultures dissolving into the global one.

The other possibility is the simultaneous flourishing of all these cultures by augmenting the conceptual framework of smaller cultures with the concepts of cooperation and peaceful coexistence. A community of constantly negotiating and cooperating local cultures would thus form a global cultural ecology. This solution entails two levels of organization, that of traditional cultures and that of a global cultural ecosystem embracing all of them. The transcendence of local cultures in this instance is the global culture, with continuous interaction between the two levels. In this case, science is entrusted with the mission of discovering ways to make this two level system functional. Scientific minds must solve the problem of the globalization of traditional cultures, that is, how to open up and transcend them, without destroying and dissolving them. Those within whom any traditional culture resides will assume the role of an interactive contributor to global culture. The emerging global ultra-mind will have a thousand faces, each representing a local-cultural module. We, the scientists, shall have the responsibility for assisting the thought processes of the global mind, by working out the paths of communication among the formerly autonomous cultures.

Notes and References

1. Csanyi, V. "Nature and Origin of Biological and Social Information." in: K. Haefner (ed.) *Evolution of Information Processing Systems*, Springer, Berlin, (1992c): pp. 257–281.
2. Csanyi, V. *Evolutionary Systems and Society: a general theory*. Duke University Press, Durham (1989a): pp. 304.

3. Csanyi, V. "Origin of Complexity and Organizational Levels During Evolution" in: Wake, D.B. and Roth, G. (eds.) *Complex Organizational Functions: Integration and Evolution in Vertebrates*, John Wiley & Sons LTD. (1989b): pp. 349–360.

4. Csanyi, V. "Individuality and the Emergence of Culture During Evolution." *World Futures* 40 (1994): 207–213.

5. Rumelhart, D.E. "Schemata: Building Blocks of Cognition." In: Spiro,R.J., Ruce, B.C. and Brewer, W.F. (eds.) *Theoretical Issues in Reading Comprehension*. Hillsdale, N.J., Erlbaum, (1980): pp. 33–58.

6. Csanyi, V. "The Brain's Models and Communication." in: Thomas A. Sebeok and Jean Umiker-Sebeok (eds.) *The Semiotic Web*, Moyton de Gruyter, Berlin (1992b): pp. 27–43.

7. Wilson, E.O. *Sociobiology*. Harvard Univ. Press, Cambridge, Mass, 1975.

8. Banyai, E.I."A social psychophysiological approach to the understanding of hypnosis: The interaction between hypnotist and subject." *Hypnos— Swedish J. of Hypnosis and Psychother. and Psychosom. Med.* 12 (1985): 186–210.

9. Banyai, E.I. "Toward a Social Psychobiological Model of Hypnosis." In *Theories of hypnosis: Current models and Perspectives*, Lynn, S.J. and Rhue, J. (Eds) Guilford Press, New York, 1992.

10. Eibl-Eibesfeldt, I. *Human Ethology*. Aldine de Gruyter, New York (1989): pp. 523–546.

11. Donald, M. *Origins of the Modern Mind*. Harvard Univ. Press (1991): pp. 413.

12. Csanyi, V. "Ethology and the Rise of the Conceptual Thoughts." In: J. Deely (ed.) *Symbolicity* University Press of America, Lanham, MD (1992a): pp. 479–484.

13. Csanyi, V. "Human evolution: Emergence of the group-self" (commentary). *Behav. Brain Sci.* 16 (1993): 755–756.

14. Csanyi, V. "The Evolution of Culture." *World Futures* 34 (1992d): 215–223.

15. Csanyi, V. "Social Creativity." *World Futures* 31 (1991): 23–31.

16. Csanyi, V. "The shift from group cohesion to idea cohesion is a major step in Cultural Evolution." *World Futures* 29 (1990): 1–8.

12

The Guidance System of Higher Mind: Implications for Science and Science Education

David Loye

That ours is a time not only of unparalleled but also of potentially transformational change is recognized by leading scientists and other scholars in many fields.[1] Curiously, in the midst of all this turmoil—and despite the mounting evidence for the need for change here too—the instrument we scientists use and our *raison d'etre*, science itself, largely remains a static entity in two fundamental regards: classifications and prioritizing.

Perhaps the most glaring example of the problem is the increasingly evident need for the development, classification and prioritizing of a science of morality. On one hand are those who, true to a prevailing paradigm of objectivity, see no place in science for the verboten "value judgment." On the other hand are those who, seriously concerned about the rapidly widening gap between our moral and our technological evolution, are beginning to worry about the prospects for species extinction. Many recognize that something is needed of considerably more weight and impact than the present "moral studies," which are pretty much peripheral for psychology, sociology, and anthropology. Thus, whether or not it is expressed this way, there is this recognition of a need for a more adequate classification and prioritizing. But as to what to do or how to do it there is not only no consensus but not even any grounding for consensus.

The same situation prevails in at least three other areas of disjunction between the demands, needs, and urgencies of this time of
great change and the classifications and prioritizing of science. Driving the development of science from its beginning has been the
desire to be able to better predict what will happen if one discovers
or does this versus that. This desire writ large has become one of the
most pressing demands of our time as the question of the future for
humanity and the assessment of favorable versus unfavorable paths
becomes ever more fateful.[2] Similarly closely linked, is the question
of the nature of evolutionary processes and how—as at the human
level we have achieved this power—to intervene in evolution to bias
it in favorable versus unfavorable directions.[3] And third, not just
interlinked but interlocked with moral, futures, and evolutionary
concerns, is the question of action—or how do we considerably up
the efficiency and effectiveness of science in its basic function of
service to humanity.[4]

Yet in all these areas we find the same situation. As with "moral
studies," the futures concern struggles for scientific grounding and
status in a peripheral field called "futures studies." The evolutionary concern is parceled out between the "hard" micro-ethos of
biology and the "soft" macro-ethos of general evolutionary theory
in an amorphous "evolutionary studies." As for the action concern, most disparate of all, this critical component in the determination of human advancement or retrogression is reduced to a
sub-field of sub-fields called "management studies."

What can be done about this situation in which classification
and prioritizing have become a matter of critical importance bear-
ing not only on our present and immediate future but also on our
potential destiny?

Within the field of brain research lie findings that may begin to
provide an answer. I originally became involved with the study of
the brain in trying to establish a firm basis in physiology for a
psychology of prediction, or how we use the brain and mind to
predict the future. Such an interest rather speedily leads one not

only into the well-known area of studies of the left and right brain hemispheres but even more so into the far less generally known and more important area of frontal brain research. Here one finds clear evidence that our capacity for projecting and assessing futures, and planning in relation to these projections and assessments, heavily involves the frontal brain, or more specifically the prefrontal lobes of the neocortex.

In trying to understand the relation of prediction to brain structures, however, I encountered great difficulties in identifying a clear track or obtaining a useful gestalt because of what at the time seemed to be a number of confounding variables. In piecing my way through scores of studies of the frontal brain I kept running into a confounding intermixture of localizations and processes for our social and moral capacities, for our ability to track and access evolutionary processes, as well as a tantalizing juxtaposition of frontal areas with the so-called motor areas of the brain that are called into service when we move from thought to action.[7]

It finally sunk in on me that rather than this being a case of disparate confounding variables, what I was looking at was an *interlocking system of higher-order capacities of brain and mind*. It further became evident that within the function of the living organism operating within the challenge of its environment, this was a system specifically for the purpose of guidance. Given the fact that the frontal brain is the latest and "highest" part of the organism to emerge both in the evolution of species and in the development of the individual as an embryo, I decided to call this entity the *guidance system of higher mind*.

Down the graphic in the center and in left hand column, under the heading *Brain Information Processing Stages*, Figure 12.1 shows the structure and sequence for this guidance system as it seems to operate in terms of the processing of information by an organism from input to output phases.

In other words, if in systems analytical terms we visualize the organism as not only a processor of food but even more importantly

THE GUIDANCE SYSTEM OF HIGHER MIND
A Comparison of Brain Information Processing Stages and an
Expanded Methodology for Science

BRAIN INFORMATION PROCESSING STAGES		SCIENTIFIC METHODOLOGY
	INPUT	
Systems Sensitivity		Empirical–Analytical Science
Social Sensitivity		Historical–Hermeneutic Science
Futures Sensitivity		Futures Science
Moral Sensitivity		Moral (or Normative) Science
Evolutionary Sensitivity		Evolutionary Science
Managerial Sensitivity		Action Science
	OUTPUT	

Figure 12.1

as a processor of information, a sequence of six information-proc-
essing stages or "sensitivities" becomes apparent. This sequence
begins with something happening—a sudden event in the world
external to us "out there," let us say. From input to output, the
organism then seems to evaluate this event or object first for its
physical coordinates in terms of a "systems sensitivity" (is it large
or small, fast or slow, coming toward or going away from me,
etc.?); and then, through a "social sensitivity," for the positive or
negative valences this event or object may have for the organism
(is it potentially painful or pleasurable, food or non-food, friend
or foe, etc.?)

These two processes seem to establish for us the nature of *what is*,
or what presently exists. But then we have the higher order
sequence that in our daily lives, as throughout evolution for all
organisms, lifts us from the world of what presently exists both to
project the future that may come to be and to visualize how we
may co-create the future that is within our power of influence. In
relation to whatever has just happened, we have the assessment of
our "futures sensitivity" for possible consequences, or what this

happening can lead to, of *what can be*. We then have this key assessment for the higher mammals (e.g., dolphins and bonobo chimpanzees as well as most particularly the human being) given the capacity for doing this: our "moral sensitivity," or of what our responsive action *should be*. Beyond this we have what seems to be the wholistic reading of an "evolutionary sensitivity," or the gaining of an overall sense of the interlinking systems dynamics of what was, is, can and even should be as a time-transcendent gestalt. And finally, guiding the output of specific action, we have the decision-making capability of a "managerial sensitivity," or a reading on how to handle what has happened, or how specifically we may get from here to where, in relation to what has happened, we next need to be.[8]

Now what might this brain/mind guidance system have to do with science education—and particularly the need for an adequate science of morality? In further pondering this arresting model I was struck by the relevance of another level of meaning.

By now practically every scientist contemplating the problem of how to properly teach science must recognize the necessity for a "hands-on" understanding of *method*. The microscope unlocks the world of the microcosm. The telescope unlocks the world of the macrocosm, etc. But methodology of course involves much more than this. It specifically involves the *operational* classification and prioritizing of the various approaches to validating knowledge that are favored by and thus shape the various fields of science. Here there are many schemes. Capturing the essence of a debate over 100 years on the nature and merits of two primary approaches, the classification scheme of philosopher Jurgen Habermas sorts methodology first into the empirical-analytical approach closely linked to natural science and logical positivism, of the Cartesian X and Y coordinates, statistics, and objectivity, and secondly into the counter-positional *historical-hermeneutic* approach of the historian and others focusing on analysis of texts, contexts, interactive processes, and other matters of subjectivity.

If we consider these two historical methodological opponents carefully in relation to this new brain/mind guidance model two things become apparent. As shown by the items listed under *Scientific Methodology* in Figure 12.1, they seem to be analogues or isomorphs for the first two "sensitivities." "Systems sensitivity" has its analogue in scientific methodology in the empirical-analytical approach, and "social sensitivity" has its analogue in the historical-hermeneutic approach. But then something of deeper structural meaning becomes apparent. For as in their apparent counterparts in brain operations, the empirical-analytical and the historical-hermeneutic approaches are designed to establish the nature of what presently exists, what is—*but they do not account for the other operations upon which creativity, evolution, transcendence in all its forms depends.*

If we consider closely the apparent functions of the four "higher" sensitivities or brain processing stages we find the missing ground for a new classification and prioritizing system that above all reveals a place for and the dimensions of a science of morality within the sciences and science education.

As indicated in Figure 12.1, we see the functional unfolding as well as the rationale for classification and an advance in status for futures sensitivity in a full-fledged *futures science*, of moral sensitivity in a full-fledged *normative or moral science*, of evolutionary sensitivity in a full-fledged *evolutionary science*, and of managerial sensitivity—building upon establishment by psychologist Kurt Lewin of this as a vital category—in a full-fledged *action science*.[10]

What would such an expanded picture of methodological approaches do for science and science education? At first all this might seem rather alien and unnecessary from the viewpoint of the natural scientist. "I am involved in the empirical-analytical and that's pretty much it," the stereotype might tell himself or reason why things shouldn't just go on as before, with the rest left up to the messiness of social science regarding which the stereotype feels grateful she or he is not professionally involved. But the world beyond the microscope, telescope, or carefully delimited and controlled site in which the natural scientist is most comfortable requires the natural scientist to resonate to, provide counsel for, and participate in policy decisions that in turn require an expansion of mind and training into areas traditionally relegated to social science.

Similarly the social scientist is faced with a situation requiring a new grounding in the natural sciences. For a quick example I could cite my own situation as a social psychologist. The older and more involved I become in the world and in the need for more adequate therapies for the healing of cultures as well as individuals, the more I encounter the need for self-education in the natural sciences that I wish I had recognized earlier.

In short, if we look closely at the real-world challenges of the professional lives—rather than the book-world challenges of the educational lives—of both natural and social scientists we see that both are involved in the working reality of a crossing of the old barriers. Many of those trained in the natural sciences are heavily involved in futures, evolutionary, and managerial or action science. Many of those trained in the social sciences are heavily involved in the mathematics most closely associated with natural science. Most importantly, at the leading edge for the development of the new theories of complexity—e.g., nonlinear dynamics, chaos, and self-organizing theories—one finds the fertile intellectual cross-breeding of natural scientists, social scientists, and other scholars who roam freely across the borders to pick and choose from wherever across the spectrum of science they find what they need.[11]

So I feel that in this new classification model embracing empirical, historical, futures, moral, evolutionary, and action science—which seems to mirror this "guidance system of higher mind" structure and sequencing for perception, learning, feeling, thinking, and action laid down by evolution in the human brain—can be found a grounding, place, and rationale for the full range of concerns that face the scientist and the scientific educator.

But here also we may find the missing guide to *prioritizing*. In general, as the perspective shown in Figure 12.2 is intended to help visualize, this model helps make the case for more educational and academic status—as well as more governmental, foundation and private funding—for the sciences that probe *what can be* in contrast to our present overwhelming investment in science that seeks to establish in ever greater detail *what is*. But above all, we find here the grounding place, and rationale for the science of morality that the widening gap between our moral and our technological evolution, and the increasingly horrifying realities of our world today, make a matter of such urgency.

What I have presented here is not, for a change, just one more call for a science of morality to add to the forlorn lineage of voices crying in the wilderness that range from Comte and Durkheim in sociology into the present.[12] What we have here is what seems to be a secure placement for a science of morality within the

THE GUIDANCE SYSTEM OF HIGHER MIND
Prioritizing Mind and Science

DETERMINES THE NATURE OF WHAT IS	DETERMINES THE NATURE OF WHAT WAS, IS, CAN BE, AND SHOULD BE
Systems Sensitivity/ Empirical Science	Futures Sensitivity/ Futures Science
Social Sensitivity/ Hermeneutic Science	MORAL SENSITIVITY/ MORAL SCIENCE
	Evolutionary Sensitivity/ Evolutionary Science
	Managerial Sensitivity/ Action Science

Figure 12.2

structures and functions of the brain within evolution, and within a potentially more fully-dimensional science that uses brain to fathom evolution.

Of many implications of this new model in this regard, I think one above all provides a key not only to the creation but most importantly to the effective *use* of a new science of morality within and by the sciences. This can be seen in the fact that morality— and even more so the components my work emphasizes of moral sensitivity and moral transformation—is not in this model off by itself in a social or intellectual ghetto, as customarily it has been handled. It is not just a matter for a sermon on Sunday or for squirreling away in one more "token" course in a crowded curricula.

This model clearly indicates that moral sensitivity:

1) functions not by itself but *through its embedding within a personal and social guidance system to which it seems to contribute a key directional component for human evolution as a matter of the co-creative impact on evolution of the caring human individual and the caring human group as agents.* To some this will be immediately apparent. To others this interpretation must conflict with what to them has become holy writ for science and the case must be made at great length as I am doing elsewhere.[13]

2) *is dependent on the other components of this system for its lifeblood.* That is, moral sensitivity can function only to the extent to which we have the futures sensitivity, or imagination, to project the consequences of our actions; the evolutionary sensitivity, or sense of the interlocking dynamics of past, present and future, to feel personally and ideally responsibly involved in this movement; and the managerial sensitivity, or capacity for the will and courageous decision-making, to take action in the moral (rather than amoral or immoral) directions that this core capacity of moral sensitivity reveals.

Moreover, I think this model further reveals why it is impossible to become morally realized without an adequate functioning also of the fundamental processing stages that identify the nature of what presently exists: *systems* sensitivity and *social* sensitivity; as well as for a continuing grounding emphasis on empirical-analytic and historical-hermeneutic science. As for practically everything else, morality involves a concern with systems stability, the state of *what is* or what presently exists to which conservatives are sensitive, as well as the state of systems change, or *what can be*, to which liberals resonate. Here too in systems sensitivity and social sensitivity I feel we may glimpse a bridge across the ages between what long ago Gautama, Jesus, and other spiritual explorers of goodness forcefully expressed in religion and what I am convinced that science is now beginning to confirm: that only as our vision of the basic relationships of *all* that composes our universe and as our compassion is opened to this multiversity can we be morally transformed.

What all this suggests for science and science education is obviously not something that can be done overnight. We cannot take a pill for it and wake up "cured" in the morning. But this new basis for classification and prioritizing can give us two vital components for the expansion of science's service to humanity that our times call for: a model embracing and accounting for the wider database that our times force us to resonate to, and a transformational gestalt to provide a useful guide for scientific and educational action.

Notes and References

1. This has been recognized and written about by a long lineage including historian Henry Adams in the last century, cultural analyst Lewis Mumford, sociologist Pitirim Sorokin, economist Hazel Henderson, and futurists such as Willis Harman. Particularly notable are works by participants in the conference for which this paper has been prepared, all co-founding members of the General Evolution Research Group originated and directed by Ervin Laszlo: *Evolution* (New Science Library, 1987), *The Choice* (Tarcher, 1994), and other books by Ervin Laszlo; *The Life Era* (Atlantic Monthly Press, 1987) by Eric Chaisson; and *The Chalice and the Blade* (Harper & Row, 1987) and *Sacred Pleasure* (HarperSanFrancisco, 1995) by Riane Eisler.

2. This is a central driving concern for the field of futures studies, notably including well known studies funded by the Club of Rome as well as most of the above works.

3. Again we find a driving concern for futures studies, particularly well-articulated by Laszlo, Eisler, and Buckminster Fuller.

4. Notably the pivotal historical concern of both Marx and Engels and John Stuart Mill, within this century the emphasis on actionoriented science was most forcefully articulated by psychologist Kurt Lewin, originator of the concept of and many techniques for action research. It has also been a major concern for Laszlo, Eisler, and myself in such books as *The Healing of a Nation* (Norton, 1971), *The Leadership Passion* (Jossey-Bass, 1977), and *The Knowable Future* (Wiley, 1978).

5. See Loye, *The Sphinx and the Rainbow* (New Science Library, 1983; Bantam Books, 1984); "The Brain, The Mind, and the Future," *Technological Forecasting and Social Change*, 23 (1983): pp. 267–280.

6. See Alexander Luria, *The Working Brain* (Basic Books, 1973); Paul MacLean, *The Triune Brain in Evolution* (Plenum Press, 1990); Karl Pribram, *Brain and Perception* (Erlebaum, 1991), and many other works in the field of brain research.

7. Standard works on frontal brain research by Ward Halsread, Walter Freeman, Joaquin Fuster, and others are sources for the scores of studies that in bits and scraps reveal the data for the model I propose.

8. This complex sequence is covered elsewhere in "Moral Sensitivity and the Evolution of Higher Mind," *World Futures: The Journal of General Evolution*, 30, 1–2 (1990): pp. 41–52, and in *The Healing of Our World: A New Science of Moral Transformation*, which I am completing.

9. Jurgen Habermas, *Knowledge and Human Interests* (Beacon Press, 1981). R.J. Bernstein, *The Restructuring of Social and Political Theory* (The University of Pennsylvania Press, 1976).

10. See note 6.

11. David Loye and Riane Eisler, "Chaos and Transformation: Implications of Non-equilibrium Theory for Science and Society," *Behavioral Science*, 32 (1987): 53–65; Ervin Laszlo, *Evolution*.

12. See Robert Bellah, editor, Emile Durkheim on morality and Society (University of Chicago Press, 1973). The only person so far to make much headway in establishing such a science was psychologist Lawrence Kohlberg,

the major pioneer in this direction. However Kohlberg's effort essentially collapsed because of its inadequacies and no one has since picked up the challenge at the Kohlberg level of the creation of an *institution* in the sociological sense rather than just more "moral studies."

13. In addition to *The Healing of Our World* (note 8), I have nearly completed *The River and the Star: the Lost Story of the Great Explorers of Goodness* and *Natural Process Moral Learning and Healing*.

14. As I write of in *The Healing of Our World*, I am finding the relevance to this model of the perceptions of the great spiritual explorers of goodness of profound significance. Gautama most notably expresses the necessity for resonation to the natural world or systems sensitivity; Jesus, Gautama and all the rest the necessity for social sensitivity; the Jewish prophets and Confucius—via his editing of the Chinese forecasting classic I Ching—the necessity for future sensitivity; all, of course, the necessity for moral sensitivity; Hindiusm in the vast scheme of reincarnation, Karma and a sequencing of ages the necessity for evolutionary sensitivity; and all, again, the necessity for managerial sensitivity, ranging from the micro-action advocated by Hindiusm and Buddhism, to the macro-action advocated by Judaism, Christianity and Mohammedism.

13

A Question of Will, Not Knowledge

Janet Ward

A Personal Introduction

When I was a small child, I felt safe. Never mind that I was born during a World War or that during my second year of life a tremendous new weapon of mass destruction was detonated and the foundations of a technology to replace my limited human intelligence with that of a computing machine were laid.

Growing up I continued to feel safe, even as Watson and Crick cracked the code of life and opened the way for human beings to manipulate their own evolution.

I was too young to understand the enormous significance of these discoveries. As a grade school and even a high school student, Nature seemed to be so vast and, by and large, beneficent that I could not imagine that human exploration and tinkering, fueled by what seemed to be natural and appropriate human curiosity, could affect it in any significant way.

While swimming one day with teen-age friends at a New England beach, I remember shading my eyes and looking out toward the ocean, thinking how inconceivably rich and deep the Atlantic was. I imagined that it teemed with what surely must be an inexhaustible supply of all kinds of sea creatures.

At that moment, the earth seemed to be a nurturing and safe haven, and presumably always would be, if only we could "control" the bomb. That seemed to be the singular challenge.

Times Change

There is a vast difference between the world I was born into and grew up in and the world my nine-month old grandson has recently inherited. Trees still grow toward the sky and rain still falls to the earth, but there is grave concern about the declining health of many of those trees and the potentially harmful chemical constituents of the rain. Indeed, the overall health of Nature, the Atlantic Ocean included, seems to be in question.

Although everyone still eats food as they did when I was a child, there is now an underlying concern about the accumulation of a variety of pesticide residues in that food. The availability of food for a burgeoning world population has been called into question. Huge catches hauled in by high-tech fishing gear from the mistakenly construed "inexhaustible" ocean bounty has seriously strained the oceans' ability to replenish that bounty.

Water remains a necessity for living things, but the quantity and quality of that water in populated areas has become a matter of concern. Weather patterns, always a subject of interest for ordinary folk as well as farmers, have also become a subject of concern. There is ongoing discussion of the effect of the effluents of an industrialized world on the atmosphere, upon which we depend so completely. There is the possibility of a depletion of the protection which the atmosphere provides against harmful atmospheric rays and the sparking of damaging weather systems.

When I was growing up medical science was armed with antibiotics, a preventive for the scourge of polio, and an overwhelmingly reassuring confidence that it would eventually triumph over a panoply of diseases. Marvelous progress has been made. But, in my grandson's world, AIDS, the ebola virus, and the evolution of other drug-resistant microbes have shaken confidence in the eventual, inevitable triumph of medical science over disease. Even the beneficence of some of medical science's progress is seen as questionable.

Life expectancy in some countries has been significantly increased, resulting in rising medical costs spawning a thriving industry to cater to the needs of senior citizens and draining private and public financial resources. The care of a fast-growing senior population, coupled with medically feasible but costly medical care, has plunged the legislatures of countries like the United States into heated debate about humane and economically feasible solutions to the challenge of making all possible medical care available to its citizens.

The list of the achievements of modern science and technology is an impressive and complex one. The omnipresence of stunningly potent military technology in wars all over the world can be positioned alongside the successful repair of the redoutable astronomical explorer, the Hubble Space Telescope. The exponential increase in the speed and memory capacity of the personal computer can be considered, along with the social dislocation caused by the replacement of innumerable human information "processors" by computer chips, and automated voice recognition and response systems. Mechanization and computerization have improved productivity, in part by providing opportunities for "down-sizing" which means the elimination of thousands of jobs. Jobless people begin to wonder about the meaning of the word "progress."

Today, many of us interface mainly with machines, not with people, during the daily work and business transactions of our lives. Indeed, many of us operate mainly in a face-to-screen mode: the screen is that of a computer, a television or a movie theatre.

Studies estimate that young children in the United States spend more than twenty hours per week watching television. Reflecting on this, and factoring in time spent playing video and computer games and attending movies, we see that many hours formerly spent in outdoor play and casual but important social interaction have been lost. This raises questions about how the process of socialization will now occur, particularly if children are encouraged to spend more time at the computer screen during each school day.

To meet financial obligations, many mothers, fathers and other guardians of children must work. Given rigid schedules and understandable fatigue at the end of the work day, many adults are not available to directly influence how their children spend the greater part of each day. Much of this responsibility has been passed on to schools.

What should schools teach, both in terms of basic academic skills and content, and what attitudes toward society, the environment and the self should be fostered? Such questions have become crucial social concerns. Yet the rapid and stunning changes in the world, brought about largely by progress in science and technology, have produced much confusion among many of the adults responsible for determining curricular goals and content. Many in the educational community are hard pressed to articulate a curriculum that makes sense given the realities of our global situation which, in so many ways, is vastly different from that in which many of the decision-makers grew to adulthood. The need for productive reflection and action is urgent.

Promising New Trends in Science

Modern Science is a relatively young force among those potent influences which have the power to shape human society. Historically the roots of modern science can be found in the study of Natural Philosophy. Curiosity about the origin and nature of the world and its inhabitants inspired the pre-Socratic Greek philosophers to opine that the "stuff" out of which the world arose was essentially water or fire, earth or air, or multifarious combinations of such elements.

These original natural philosophers sought a convincing, rational explanation of how the cosmos began, and how it changes and becomes everything, including us. For many, earlier creation myths and stories were surpassed by pre-Socratic explanations which often possessed a compelling, persuasive power.

In the sixteenth century, the English philosopher Francis Bacon spoke of pushing and poking Nature, of forcing her to reveal her secrets. This crude description pointed the way to the notion of scientific experimentation like that of Galileo who devised ingenious ways of exploring the laws of motion. Much later, Louis Pasteur's own clever experimentation settled the question of life's spontaneous generation.

Over time, the scientific community has become a circle of like-minded thinkers inspired by a shared curiosity about the way the world works, and engaged in an enormously grand enterprise: the explanation of all natural phenomena.

We, too, you and I, are natural phenomena, but we are phenomena possessing a quality apparently unique to our species: subjectivity, an awareness of self. Subjectivity is difficult to locate precisely, to observe and to quantify even with the most advanced brain imaging techniques. Only recently has it become a subject of scientific inquiry. The 1990's, the decade into which my grandson has been born, have been called the Decade of the Brain. It is an era in which science is turning its powerful focus inward and examining the human brain/mind. Where will this and related scientific exploration lead? Consider this headline in *The New York Times* (August 22, 1995) "Neuron Talks to Chip and Chip to Nerve Cell—a big step toward two-way links for mind and machine."

And this item in the 150th Anniversary issue of *Scientific American* (September 1995). Arthur Kaplan, founding president of the American Association for Bioethics, stated that, because of progress in our understanding of genetic manipulation and related science and technology, "We need to decide to what extent we want *to design our descendants.*" (Emphasis mine. See p. 142, "An Improved Future?")

And there is yet another aspect of natural phenomena which science has only recently begun to explore: the "connectedness" of everything with everything else. Some refer to a "systems approach", some discuss complexity. Regardless of the terminology, there is a new *zeitgeist* developing, and it has to do with the recognition of the interrelatedness of everything in nature. There appears to be the beginning of an authentic and growing appreciation that everything is a part of a cosmic nexus.

Interdisciplinarity has become essential. Philosophy, history, literature, mathematics, psychology, the social sciences, and the arts must now make room for scientists' efforts in molecular biology, physics, neuroscience, cognitive science and complexity science to deal with questions which were once the purview of the humanities alone. What is a human being? What is a human being for? Such questions, once thought to be the province of philosophers in ivory towers, are now pressing and real ones in scientific laboratories as well.

Science and technology have made it possible to do any number of amazing things. Our recent history has made clear that all of these amazing things have consequences, sometimes unexpected and painful consequences. Using the knowledge borne of our difficult history, it is time to move forward wisely. *Now it is not a question of knowledge, it is a question of will.*

To speak of the wholly objective and amoral nature of the scientific endeavor is no longer possible. Although a particular laboratory experiment may not, in itself, possess moral significance, every scientist is a human being, a significant member of a global community. It is obtuse to suggest that scientists, and science educators, need not consider the possible results of their activities on society.

As our scientific study of nature progresses, our examination of the constituents of systems requires us to ask how those constitutents act in concert, within the whole systems of which they are a part, how they work *together* to account for a phenomenon such as my "I", or the weather, or the ecological "health" or "illness" of our global home.

To pose the need to "decide to what extent we want to design our descendants" is to plunge our global society into an examination of profound philosophical issues, issues which perhaps Francis Bacon could ignore, but we cannot. We do not have the luxury of ignorance. We have seen the atomic fireball, if not in person, then on film. Some of us have been poisoned by toxic wastes. Many have felt the sting of unusually severe weather systems. Many of us now purchase bottled water because we feel we cannot rely on the supply and purity of water from local sources. We now pay dearly for once inexpensive and plentiful seafood. All of us, in some way, have felt the unintended consequences of science and technology's so-called "amoral" actions. Now we must act on what we have learned.

School—A Logical Locus of Change

We have learned that science and its concomitant technology is responsible for many of the challenges facing the global community today. Since science is taught primarily in the schools, how should it be taught to insure that science and technology are responsibly used in the future? What should we teach future generations, my grandson's among them, so that they can have reason, despite the many ominous trends which science and technology have set in motion, to hope for a happier future?

I believe curricular content need not change radically. Rather, the following fundamental concepts and attitudes, a set of new habits of thought, must be incorporated and taught explicitly, in developmentally appropriate ways, and modelled by teachers in every science classroom, indeed in every classroom, from kindergarten through post-graduate studies:

- We human beings are not independent organisms. We are organisms dependent for our survival on our environment.
- Human beings are an integral and vital part of nature. (By nature, I mean the context in which we exist, that nexus of matter and energy, of which we are an integral part.)
- We human beings have developed the power to bring about vast changes in our physical environment by bringing about vast changes in our *intellectual* environment. Francis Bacon's

attitude toward nature was and is not a universal one. It was constructed by Francis Bacon and adopted by thinkers who went forward on the path he recommended to them. Other beliefs about nature and the relation of our species to the natural world would have elicited different actions, perhaps more respectful, more cautious, less damaging actions.

- All human history is a construct. It is true that our environment impacts us, but so do we impact our environment.

Our physical impact on our environment is shaped and inspired by *how we think* about our environment. Our knowledge of the unintended damage past "progress" in science and technology has wrought can inspire us to use our tremendous intellectual and physical powers in life-sustaining, life-enhancing ways. We have adequate knowledge in hand now to begin to build a better world for future generations. A better world is possible. We have only to choose to *act* on that knowledge.

14

Some Remarks about Education

Willem Brouwer

It is clear that, at least in the United States of America, education rarely approaches expectations. Of course, this means students are not well equipped to pursue careers and are unprepared to enter institutes of higher learning. In this essay I would like to identify some major problems besetting school systems and propose ways to improve the situation; these remarks include home teaching and the providing of tools to make the process easier for parents and teachers alike. One such type of tool might be well thought-out CD-ROMs to help teach the necessary subjects in ways designed to respect and preserve the child's creativity.

It appears that a major problem in formulating solutions is that many classrooms simply have too many pupils for teachers to give enough individual attention to each child. Fiscal and political problems prevent the building of more schools and the hiring of more teachers. Many teachers also claim that they need more cooperation of parents, while parents feel that they are not welcome in the teaching process, although many of them would love to help in the education of their children. Their cooperation is absolutely essential.

The Cambridge-based Holt Associates, a clearing house of information about home schooling, estimates that in Massachusetts today more than 2,500 children are home-schooled. Certainly this shows that there are parents who are willing to go to great lengths to educate their children. Such groups, with or without involvement with public schools, could do a great deal to help improve the children's education.

Since a great deal of learning has already taken place before the child enters school, I propose to examine how this learning (education) is accomplished in preschool days. Most children learn an enormous amount! They manage to grasp their own reality while forming a picture of the outside world, both in space and in time. Not only that, they learn to cope with people and to judge their intentions, and to recognize and avoid dangers. Most of this is acquired without the intentional participation of the parents other than their being supportive, loving and serving as an example. The amount of learning in these first few years is truly mind-boggling. All children are extremely interested in and curious about their surroundings. Children are typically also very persistent in learning to do all the things they want to.

Let me give you a living example. It involves one of my own children, when he was between three and four years old and learning to play with a top. He had a little whip and would wind it around his top. He then placed it in a hole he had made in our lawn. He jerked his whip to start it spinning and then tried to keep the top going by repeatedly hitting it. Of course he failed every time. This was no reason to give up and he would continue to make new holes and try again. Since the yard was my wife's pride (we lived in Holland at the time) and she did not want all the holes in the lawn, she forbade him to play with his toy in the grass. I was sitting in a chair on the cement tile covered area near the back door and was curious how he would respond to this limitation of his freedom. Well, he sat down for a while and then disappeared, returning with a new head for a broom. On the two slanted edges were holes to insert the broom handle. He propped up this head so he could insert the top in one of the holes and proceeded to learn to work the top, this time with much more success since the platform was a much better surface to spin the top. This is a good example of what I am talking about, children naturally exhibiting creative thinking and persistence in their pursuits. Almost all children display these characteristics! As a matter of fact, if they don't we conclude that something is quite wrong and take them to the doctor or psychiatrist!

To summarize this phase of life, it is a period of enormous learning, done without specific "teaching" by outsiders. It is done by the child with great creativity while he/she forms fundamental ideas about reality. There seems to be no slowing down as entry to school approaches; there must be a reason why this enthusiastic drive to learn should diminish or, occasionally, actually cease, when he enters a school. There must be a fundamental reason. Are teaching methods in the schools the culprits and are they at least partially responsible for this? I believe they are and the reasons are of evolutionary origin.

Before speaking had evolved, *Homo* already had the neuronal system to form an internal understanding of the outside reality, formed by the inputs of his senses, and coded in the synaptic connections in the brain. It was much later that humankind started to attach words to such internal ideas and to recall the meaning when the word was used. Contrast this with the revolutionary concept of first learning a new word and then acquiring the meaning by the development of the neuronal pathways to construct and store the symbolic representation in neuronal form. This entirely *new* process must be learned from scratch. So when a child goes to school, and is taught by a methodology in which prior experience is no longer the basis for relating the word to its understanding, we should expect the child to be unprepared for this reverse process. Or expressed in a different way, learning by oneself is an entirely different process than being taught. It is therefore not surprising, especially in the US where the children are of such diverse background, that many of them have great problems with this abrupt change of their method of learning! This is especially the case when entering the world of abstract meanings in math and science. We shouldn't be surprised that many kids get completely lost and never adequately accommodate to this process. No special attention is paid to make sure that he/she learns this new process. We must make this process more natural and gradual.

Extreme care should be used in making sure that the understanding can be based on words the child already knows the meaning of. Only those curious enough to want to learn will ask questions to find the missing parts of their understanding. The opportunity to get these answers should always be available. Otherwise the new words only become meaningless "buzzwords."

The experiences I had during the war in Holland were at the free school started by Kees Boeke based on the model of the Summerhill School initiated by A.S. Neill in England. In these schools kids do not have to do anything except two things: first, they have to brush their teeth every night and, second, they have to work in the vegetable garden for two hours every week if they want to eat. Furthermore, the kids are the final judges and are the only ones that can hand out punishment. Should you think that only children of very progressive parents attended this school you could not be more wrong. In actuality, since they could not give a diploma but only a promise that they would *try* to get their child enrolled in a high school, this was a guarantee that only parents at the end of their rope, after all else failed, would send their kids there. So the typical student that arrived had been thrown out of other schools due to misbehavior. They all started by doing nothing of course! All the other kids were, however, busy pursuing their interests and did not have time to "hang around" with the neophytes. In short order the newcomer discovered that all destructive acts would be severely punished by the kids' tribunal. This was especially true here since the school was maintained, and, if necessary, enlarged by the kids themselves. As a result they all took great pride in their school and tolerated no behavior that would diminish the respect the school had in the neighborhood! Admittedly the English school was not originally constructed by students, but Neill reports that the damage done by new pupils was negligible. As it was, the newcomer would soon be bored and, within a short time, find some group to join and start learning with this group. Each was free to join more than one group and leave any group in which they lost interest. The end result was that students "graduating" from the school were invariably accepted at the finest high schools in the country!

How did these kids learn enough to become well-educated individuals? Well, let me give a hypothetical example of the career of such a kid. Suppose he/she was only interested, let us say, to take some very unusual activity, ballet dancing. At my time there was no teacher to give ballet lessons, but they would probably convince, for example, the physical education teacher (or any other for that matter) to help them get started. He would generally suggest the student find at least a couple of others who would be interested since dancing would be much more fun with a group of people. She or he would usually be back within half an hour with the required comrades. In the meantime, the teacher would have probably found a number of book titles to get a start. He would help the dancers as well as he could and maybe find an acquaintance who was willing to help. You would be amazed how fast the kids read all the books they could get hold of. Soon they would be back for more help. Here is where "educational leverage" comes in. The teacher might ask them if they could read French. Let us say their answers were "no." "Too bad, because if you could read French there are some magnificent books you could read!" Again they would start to learn French with some help and do this in a surprisingly short time. Or the teacher (guide) could say: "too bad that you do not know mechanics, a part of physics, because proper dancing is often a question of dynamics and equilibrium." Off the kids would go to the physics teacher! He would not give them a course, but he would tell them to read a book and to come ask questions if they could not understand it when they really tried. I saw scenarios like this one performed again and again. By this means eventually all the kids got a great education which was probably broader than that achieved by the typical regular school at that time, and certainly far better than most kids in present day American schools. The real trick was that they were not forced to learn. Yet whenever they wanted, means for learning were always available to them, in a way that usually was very much connected with their reality.

What I learned from these experiences was that any normal kid can learn quickly and in depth when he *wants to!* This learning is closely connected with curiosity and desire to know; such interest should be kindled first. This process is, at least, part of the real secret; he cannot be forced to learn, probably not in a large group, on command of the teacher and in blocks of one hour to boot. It turns most minds off.

I can almost hear you saying, "Wait a minute, this will lead to total chaos and will take too much time," and much more. I agree that it is impossible within the present school system. This suggests that we should try to find another solution to bring us out of this trap. The real question is: how do we go about teaching kids so they stay interested, retain natural creativity, and learn all that they can absorb? This is a really difficult problem but I do believe that we can make a few constructive comments about it that are helpful.

There is another group of people with a vital interest in the well being of children, their children. I am, of course, talking about the parents, many of whom are deeply worried and disgusted with the present system. A good indicator for this is the success groups have that advocate and supply materials to help children learn reading using the old-fashioned methods of phonics. Other groups organize small circles that cooperate in educating their children at home. There should be a way to bring these groups together with the schools to cooperate and at the same time greatly improve the transition from self-learning to being taught.

From my own experiences in school, I know the great importance of the few teachers who took an interest in me and challenged me and helped me all along. I greatly admired these and they became important mentors to me. I was lucky to have so many of them. Later, during my twenty-one years of being a boy scout leader, I asked all the kids in my patrols if they had a special teacher they admired. Only a few had and often this was a sports teacher. This was probably true since the sports teacher demanded performance and challenged them! These two points seem especially important to me: having a mentor and being challenged.

From my experience as a boy scout leader I also learned that parents cannot usually be the child's hero. Parents have to keep telling them to shut the door behind them and eat properly and brush their teeth. On top of that parents are too heavily involved emotionally. Children need non-parent adult mentors. This means that parents who want to teach their own children should form groups of families and do the job cooperatively. This gives the kids a chance to work with other adults and the opportunity to discover mentors they really admire. The second fact is that children need to be challenged. Nothing is more destructive than being bored and it is a well-proven fact that kids can learn a lot and at a furious pace. For instance, two languages can be learned simultaneously as fast as one. We need not worry about teaching them too much.

In their early years, children are incredibly curious about their surroundings and are great observers. Reality is always fascinating. I would suggest that parents and schools recognize that it is important for children to see what is going on in the real world. The parent groups could again play an important role. They should invite the children to visit their mother's and father's work places. And, not only show them what it is that is done, but what their own job is and how it fits in the total picture. Many businesses will organize tours to show their plants. Again try to visit as many of those as you can. The greater the variety the better. Emphasize again what the role of people is everywhere. How can kids select a career if they do not know what is available in the real world?

Support for the groups that encourage these home learning groups and schools would be a valuable way of encouraging this type of education. Beside general information on teaching practices, development of appropriate teaching aids would be invaluable, especially teaching aids that encourage creative learning. It might be possible to use some newer computer methods for this purpose, especially cleverly designed CD-ROMs for teaching specific subjects. I am not thinking of the usual methods of making teaching into a game of simple information packages. Techniques should be developed where the child is challenged to think creatively about the subject. For instance, instead of giving the information it might ask the student leading questions to be answered before correct answers are given, or new questions based on the answers that might show the fallacy of given answers. The teachings should be all inclusive. For instance, in a course of physics all the formulae and facts should be taught, relevant to the age of the kids. It might, however, be better to push too hard than to artificially make the subject appear simple. Again kids like to be challenged and be proud of their accomplishments. I remember a math teacher who always came to me during a test and told me in how many steps he could solve the problems (the kind that a statement had to be proven and ends with q.e.d.). It turned out that I could always do at least one better and I always felt real proud if I could do one or more in fewer steps!

I am convinced that it is possible to bring these groups together and that methods of such cooperation must be studied and experimented with. This is the path to improvement in education necessary in our modern world. An all out effort to create these programs will have surprising results.

15

Science Literacy for the 21st Century:
The Role of Science Centers[1]
David Ellis

As the world has become more and more technologically oriented, it has become ever more clear that our society needs not only well-trained scientists and engineers, but also workers who are sufficiently knowledgeable to undertake a wide variety of jobs requiring some basic understanding of science and technology without becoming professionals in those fields, and a generally scientifically literate populace as well.

Cognizant of the numerous studies in recent years that have shown that young people in the United States are not well prepared in science and mathematics [2,3,4,5,6]—our students' accomplishments do not compare favorably to those of comparable grade level students in other countries, both developed and developing[7]—and that a quite low percentage of the general populace can be described as scientifically literate, [8,9,10] those of us who are concerned about science education are faced with bringing to bear effective strategies for this situation. In my current role at the Museum of Science, I have been able to observe that science centers are important contributors in unique ways to the struggle for science literacy and can contribute more in the future.

Science Centers Today

Science centers today contribute significantly to science education and scientific literacy through experiences involving both permanent and temporary exhibits and myriads of programs, such as "on-the-floor" demonstrations, interpretation and science theater; large-screen theaters and planetariums; courses and outreach programs of many different types; leased or sold science kits for teachers or families; and the list goes on. Science centers also utilize many different display technologies and media in order to expand the educational potential to many different kinds of audiences. Science centers often become laboratories for schools and providers of pre-service and in-service training for teachers, particularly elementary school teachers.

Learning in science centers can be conceptualized in many ways. Under the tutelage of our Vice President for Exhibits, Larry Bell,[11] the Museum of Science has approached exhibit development based on the proposition that learning in science centers takes place predominately in three dimensions: knowledge, attitudes and skills.

Knowledge

A science center's special domain is experiential and non-verbal knowledge. The exposure to non-verbal knowing that can be found in a science center may take many forms; for example, seeing an object for the first time and relating it to what one may have studied or to some other object that has been seen; for example, comparing size, color, shape, texture, motion, etc.; or experimenting with an interactive exhibit. Since understanding is achieved through the interplay between the concrete knowledge gained through sensory experience ("seeing is believing") and the abstract knowledge gained through language and other symbolic forms of knowing, this kind of learning is fundamental to understanding the physical world. If science teaching is focused almost exclusively on verbal and mathematical knowledge, students may be able to repeat memorized facts but they also may be unable to apply their knowledge and understanding to new situations.

The chance to look at, study, listen to, smell, taste, touch, manipulate, operate and experiment with real things is a valuable form of conveying information and understanding. Science centers today attempt to provide these kinds of learning experiences in a motivating context. When students, children and adults find something in a science center that they have heard about but have never seen or otherwise experienced, they acquire the concrete experiences that together with abstract representations make understanding possible. It is this which often produces the "aha" experience of which museum people often speak.

Attitudes

The principal attitudes that science centers strive to promote are a joy in understanding and an interest in learning more. Active participation and entertainment in science center exhibits and programs are keys to the science center's success in developing these attitudes. Testimony to the value of the enthusiasm generated in science centers is found in anecdotal accounts of adult scientists who first became interested in science as the result of a visit to a science center. It is also found in the accounts of both planned and unplanned class projects that have followed school group visits.

Sometimes casual observers of the enthusiasm and energy evidenced by children in a science center environment question what is occurring; in this context it is important to remember that enjoyment and learning are not mutually exclusive. We think it is a compliment to refer to science centers as amusement parks for the mind; the sense of play is an important component to creativity and positive attitudes.

Skills

An old Chinese proverb says: "I hear and I forget, I see and I remember, I do and I understand". Visitors in science centers bring with them many learning skills, and as they explore different exhibits or programs, they naturally use these skills. With the increasing emphasis on interactive exhibits, science centers encourage this. The insider term "hands-on" is used to signify that science centers are fundamentally different from institutions where "hands-off" exhibits are the norm; it is a short-hand way to emphasize that visitors are encouraged to be active participants in the experience. To further the development of scientific thinking skills, science centers organize and display materials so as to make objects and materials available in a manner that encourages a broader and more open-ended exploration. The increased emphasis on the habitat and the introduction of "hands-on" opportunities in zoos and aquariums are further important example of this. In science centers, visitors often find they are asking new questions, challenging their preconceived notions, and in general developing enhanced scientific thinking skills.

Science Centers in the Future

There are several trends in the planning and management of science centers that are intended to increase their utilization in the coming decades. Among the more important of these, as I see it, are:

- stronger efforts to understand the audience being served;
- greater emphasis on the applications of science, discoveries in science, and relevance in general;
- an increased emphasis on finding ways to leverage the science center's impact through partnerships; and
- a greatly increased use of communication technology, including multi-media, wherein the technology will be the means of presentation and in other cases the subject of the exhibit.

Understanding the Audience

Science centers will serve a much broader spectrum of the public. This has begun with numerous diversity and outreach programs. The exhibits and programs offered will be designed to be relevant to this variety of prospective visitors. A spectrum of learning styles will be served. Programs for specific audiences will become more common. Both market research and front-end research will be used extensively. And very importantly, more educational research will be done to evaluate the visitors' experience, both from an educational standpoint and in terms of the overall experience.

Focusing on the audience will have other implications. Since attendance at science centers is voluntary, the real world need to compete successfully for discretionary time and money cannot be minimized. Quality service in all aspects of the science center's operation will be paramount, whether in terms of friendly, helpful and informed staff, parking, food, ticketing, places to sit, clean restrooms, well-maintained facilities, etc.

An Emphasis on Relevance: Current Discoveries and Applications of Science

As science centers attempt to better serve their publics, relevance is a key ingredient. Working to explain the modern world as we experience it must be added to the traditional explanations of the traditional fields of science. With rapid advances in knowledge and the ever increasing applications of science, coupled with a greatly increased interest, even demand in some cases, on the part of the public to understand the implications of a discovery or new application, science centers will be called upon increasingly to help address the public's understanding of science. Science centers are in many ways well situated to meet this need.

Increased Presence of Technologies, including Multimedia

Museums, zoos and aquariums display real objects. Over the last 30 years, this tradition has been supplemented in science centers of all types by interactive, "hands-on", exhibits, "theater" in the halls, and the increasing use of various types of media. This trend will undoubtedly continue. Technologies appear in science centers, however, in two distinctly different ways. Sometimes the technology is the object of the exhibit or program while in other instances it is simply part of the delivery system and in some cases it may be both. It is important to recognize this difference.

The inclusion of technologies as the objects of exhibits and programs is an important development for some science centers. Newer technologies are often of significant interest to the public; people want to understand what is happening, how it works, the implications for them, for society, for the environment, etc. There are many examples of technological objects in which some members of the public have an interest: how does a microwave oven work; how does a digital watch work; how does a radar detector work; how does a CD-ROM work; how can so much information be placed on a computer disc; how does an electronic ignition in an automobile work and how is it different from what was used before; how do the media create special effects, etc.

These newer technologies also permit the presentation of scientific concepts that were either very difficult or impossible to present earlier. A good example is the simulation of a chemical reaction which cannot be "seen" or easily demonstrated in many cases; another example would be the simulation of different weather patterns in different seasons or as influenced by an "El Nino". Increasingly visitors will be able to undertake "interactive" simulations, often using data from distant sources via the Internet, which would not have been possible previously; and very importantly, the approach to a particular subject can be drawn from a much broader range of exhibit or program methodologies, thereby better meeting the different learning styles of visitors.

With the greatly enhanced computer power of the last decade and other newer technologies, immense amounts of information can be made available through the use of CD-ROMs, laser discs, etc. Science centers are in a highly advantageous position to link these information sources to real things and to help visitors see the relevance of what is presented. The Internet will increasingly become the path to many databases, and science centers will use the ready availability of that information in many ways to serve their visitors.

The development of the Internet also permits science centers to reach out much more broadly, in terms of audiences. And it opens the possibility that to some degree what has traditionally only been available within a science center can now be available to anyone with a computer and a modem. As science centers digitize exhibits and collections, and learn how to effectively present interactive exhibits, demonstrations and programs over the Internet, the opportunity to take the strength of the science center to the people increases exponentially.

The use of these new media brings novel educational challenges. How can science centers develop programs and exhibits which use each technology in an appropriate way so as to maximize the visitors' learning and experience? There is a considerable body of largely anecdotal evidence to support the position that the "real thing" has a special power of its own. It is for this reason that decisions about whether to use the representation or the real must be carefully considered and the final decision will need to be based on what is intended to be taught, to whom and how. What are the unique values that the new technologies add? What must be avoided? How can interactivity be developed to operate effectively over the Internet? There are indeed a myriad of opportunities, coupled with nearly unlimited uncertainty. It will be essential for science centers to carefully explore what they can do and do well to maximally utilize the evolving potential of communication technologies.

Leveraging the Science Center

In order to achieve the goal of significantly enhancing science literacy, science centers are becoming partners with many other enterprises, including schools, colleges, universities, corporate laboratories, governmental laboratories, research institutes, etc.

Some examples here would perhaps be helpful. Science centers will serve increasingly as laboratories for schools and in some cases classes are taught in science centers on a regular basis.[12,13] After school and weekend programs for students and for families repre-sent an area of great opportunity, especially where they could be held in community centers, schools, etc.

Science centers have become the organizing force behind the establishment of "magnet schools."[14,15] In some states, community or charter schools established as part of education reform efforts may wish to establish relationships with science centers. Curricular materials developed by science centers with appropriate advice from teachers are currently distributed for the "home school" market or for teachers, with appropriate advice from teachers.

Students with the help of faculty in universities and colleges may develop exhibits or programs to be placed in science centers to dem-onstrate current discoveries or applications of science. Corporations may develop an exhibit in collaboration with a science center to illustrate some new technological product. Corporations, research institutes or universities may sponsor lecture series. Corporations may sponsor summer "camps". The possibilities for cooperation and leveraging are almost endless.

Serving Specialized Audiences in the Future

As science centers work to promote science literacy, they will focus more particularly on their different audiences and their par-ticular interests. Outlined below are some of the major audience groupings served by science centers.

Families with children constitute the largest single audience group for most science centers. The potential for contributing even more to science literacy with this group is great. The two major chal-lenges today are in being competitive with the many alternative attractions available to families and in attracting families who have not traditionally attended science centers. Different cultural pat-terns as well as shortages of time and funds often make this latter challenge particularly difficult; making exhibits and programs more interactive and relevant will help.

School groups constitute a second major source of attendance and an important opportunity to provide an enhanced contribution to science literacy. Formal schooling is a major contributor to science literacy (Miller, 1994) and for many schools, science center trips and curricular materials are an integral part of their programs. School visits to science centers, often called field trips, must be well planned to harvest the many potential benefits possible from such an experience for the students. The potential for true learning partnerships in this area is immense.

Teachers have, for many years, represented another important avenue whereby science centers contribute to science literacy. This is especially true for elementary school teachers. Science centers are well known and respected for their workshops, both in-service and pre-service, in helping teachers learn how to better teach in a hands-on, object-based, discovery mode. The potential to significantly increase science literacy may be greatest, in fact, by working with teachers to build upon the natural curiosity of children.

Adults without children represent a fourth category of audience. In some institutions, this group is a large majority, while in others, it is a small minority. Regardless, adults by themselves represent potentially a much larger market than currently seen in our science centers and if science centers do some of the things mentioned in the earlier sections, more adults will participate. The success of science-based large-format films (IMAX, for example) is a good example since these theaters and films often attract people who would not otherwise come to a science center. However, our goal must be to make science centers even more into places where the public learns about science, technology and their implications. Science centers often serve as "neutral ground" and should be used as places where scientists, engineers, members of the press, politicians, and other interested parties can discuss and debate issues of importance, whether on the environment, health or technological aspects of our economic well-being.

The press is a special audience group for whom the science center can be an important resource. Currently, members of the press often use science centers to obtain information or to obtain a better understanding of something that has been recently discovered.

Science centers will do even more of this and will become known as the preferred place for members of the press to seek information. Science centers also have expertise in presenting science in a way that the public can understand and they can provide a significant service to members of the press or other persons interested in enhancing their skills in this crucial area. Science centers could also develop specialized training for media personnel to understand their potential and responsibility to help foster science literacy.

Conclusion

Natural history museums, aquariums, zoos, science museums, etc. and more recently science centers and children's museums have contributed for decades to science literacy and to encouraging young people to consider careers in science and technology. In the future these institutions will contribute much more significantly to the education of young people as well as to science literacy for adults.

Notes and References

1. The term "science center" will be used to speak of all institutions that contribute to informal science education for children and adults, including especially science centers, science museums, aquariums, planetariums, zoos, museums of natural history, children's museums, etc.

2. *Educating Americans for the 21st Century: A report to the American People and the National Science Board*, National Science Board Commission on Precollege Education in Mathematics, Science and Technology, Washington D.C. 1983.

3. *A Nation at Risk: The Imperative for Educational Reform*, The National Commission on Excellence in Education, Washington D.C., 1983.

4. Scientific Literacy (entire issue), *Daedalus.* Issued as Volume 112, No. 2, Spring, of the Proceedings of the American Academy of Arts and Sciences, Cambridge, MA. 1983.

5. *The Science Report Card, National Assessment of Educational Progress*, Educational Testing Service, Princeton, N.J. 1988.

6. *Science for All Americans*, American Association for the Advancement of Science, Washington D.C. 1989.

7. International Association for the Evaluation of Educational Achievement.

8. *Science Indicators*, National Science Board, Washington D.C. 1979, 1981, 1985, 1988, 1990 and 1992.

9. Miller, D. Jon, in *Communicating Science to the Public*, John Wiley & Sons, New York, 1987.

10. Miller, D. Jon, *Scientific Literacy and Citizenship in the 21st Century*, The International Center for the Advancement of Scientific Literacy, The Chicago Academy of Sciences, Chicago, IL, 1994.

11. Bell, Larry, "The Role of Exhibits at the Museum of Science in Boston', 1992 and 'Learning Science', Museum of Science, Boston, MA. 1993.

12. *Youth Alive Program*, Museum of Science, Boston, MA.

13. Buffalo Museum of Science, Buffalo, N.Y.

14. California Museum of Science and Industry, Los Angeles, CA.

15. Science Museum of Minnesota, St. Paul, MN.

16

Networking, Interdisciplinarity, and Scientific/Technical Literacy: Perspectives from the Space Program

E. Julius Dasch

"The greatest sin committed by formal education is the continued and relentless extermination of the natural pleasure of learning; and the greatest sin committed by 'progressive' education is the delusion that such pleasure need not be accompanied by a certain measure of pain".

—Sydney Harris
Thoughts at Large

Addressing the fundamental problem of inadequate scientific and technical literacy requires a long term, concerted, and costly commitment of citizens in government, education, and industry, as well as many others fields. However, such dramatic measures must be undertaken, and made effective, for the continued growth of all people and their standards of life.

This paper reflects current thinking of personnel in the NASA National Space Grant Program on how to implement scientific and technical literacy. Space, along with such topics as oceans, ghosts, dinosaurs, and video games, are of widespread interest to youngsters of all ages, and therefore offer intrinsic vehicles for enhancing scientific as well as technical literacy. A second theme outlined herein is the role that interdisciplinarity, especially at popular and introductory levels, plays in sowing and maintaining literacy.

Space education enhancement and reform is best coordinated and accomplished through an interactive network of public and private educational institutions, space and related industry, federal, state, and local government, and nonprofit institutions. The maze of educational initiatives that are continually being developed in and among these institutions can benefit greatly from a concerted effort to evaluate, enhance, and disseminate the best and most useful.

The NASA Space Grant Program

The Space Grant Program, with broad goals in teaching, research, and public service, an example of a program described above, has recently concluded its first five years of operation. Organized as a network among the types of institutions listed above, Space Grant has initiated a multipronged program to influence space education through networking. Evaluation of the 52 state programs and the national program indicates that networking has been mainly successful in the state consortia and that it constitutes a powerful mechanism for positive change.

Background

The National Space Grant College and Fellowship Program was initiated by the U.S. Congress with the passage, on October 30, 1987, of the National Space Grant College and Fellowship Act. This act is part of the 1988 National Aeronautics and Space Administration (NASA) Authorization bill. Modeled after the U.S. Dept. of Agriculture's 130 year-old Land Grant College program, and the U.S. Dept. of Commerce-National Oceanic and Atmospheric Administration's (NOAA) 30 year-old Sea Grant program, the Space Grant program was brought about primarily as a result of the efforts of then-Senator Lloyd Bentsen, Democrat from Texas, to respond to what he termed the need for a coordinated effort to help maintain America's preeminence in aerospace science and technology.

The Act—Public Law 100–147— cited broad objectives such as assuring the economic vitality of the United States, and the quality of life of its citizens through the understanding, assessment, development and utilization of space resources. In addition, the law held that research and development of space science, space technology and space commercialization would contribute to the quality of life, to national security and to the enhancement of commerce. In recognizing these objectives, Congress urged a "broad commitment and intense involvement on the part of the Federal government in partnership with state and local governments, private industry, universities, organizations, and individuals concerned with the exploration and utilization of space..."

Objectives

To translate the objectives of the legislation into realistic and achievable goals, NASA personnel charged with administering the new program brought a number of individuals representing professional education and other associations into discussions on how best to structure the Space Grant program, and to outline program goals and the means by which to fulfill them. As a result of these discussions, the following program objectives were developed: (1) to establish a national network of universities with interests and capabilities in aeronautics, space and related fields; (2) to encourage cooperative programs among universities, aerospace industry and Federal, state and local government; (3) to enable the development of interdisciplinary education, research infrastructure and public service programs related to aeronautics, space science and technology; (4) to recruit and train professionals, especially women, underrepresented minorities and persons with disabilities, for careers in aeronautics and space-related science and engineering; and, (5) to develop a strong science, mathematics and technology education base from elementary through university levels.

Space Grant Today

Implementation of the Space Grant program resulted in the formation of a national network of organizations consisting of public and private colleges and universities with varying degrees of aeronautics and space-related resources and capabilities, industry, state and local governments, and nonprofit organizations. Space Grant Consortia are established in all 50 states, the District of Columbia, and the Commonwealth of Puerto Rico. Each consortium receives NASA funds to be used in implementing a balanced program of research infrastructure, higher education and public service, including precollege outreach. Consortia must obtain matching funds from nonfederal sources and must promote the Space Grant program throughout their respective states. Activities vary by consortium and include new undergraduate and graduate interdisciplinary academic programs and courses, research experiences for undergraduates and high school students, interinstitutional research collaborations, space resource centers, space education conferences, and precollege programs in curriculum development, career advising and teacher training.

Networking

By the end of 1994, the Space Grant program consisted of 552 participants. Colleges and universities number 395, and include two- and four-year colleges, medical and law schools, and M.S. and Ph.D. degree-granting institutions. Community colleges are represented with 60 taking part in program activities. Other partners include 26 state and local government offices, 66 industrial affiliates, 40 nonprofit organizations, and 25 described as other education entities (these can include local school districts, for example). Of the colleges and universities that have joined the Space Grant network, 38 are Historically Black Colleges/Universities; 10 are Other Minority Universities, and nine are Hispanic-Serving Institutions. Five tribal colleges, five institutions serving primarily women, one academic institution for persons with disabilities, and 25 minority-focused organizations, complete the categories of higher education institutions that form the foundation of the Space Grant network.

The development of collaborations, cooperative activities, and other types of partnerships have been emphasized since the program's earliest days. The Space Grant program invariably has been regarded as a "seed money" program only, not structured to provide increasing sums of money to its grantees. Leveraging has been underscored, not only to aid the schools in providing the requisite amount of matching funds, but to enable partners to pursue joint projects and activities that are cost-effective while benefiting all participants.

In 1994, an astonishing number of collaborative efforts in research, education and outreach were reported by the consortia, including nearly 830 described as being in the same college/university department, with over 770 occurring in other departments within the same institution. Partnerships involving other institutions of higher education numbered some 540, with another 120 involving community colleges. Industrial collaborations accounted for more than 340 instances of cooperation, while some 250 partnerships with nonprofit organizations were noted. Joint projects with NASA Centers occurred just over 400 times, and NASA Teacher Resource Centers were cited just over 200 times. Other Federal government agencies participated in nearly 220 projects, while organizations representing women, underrepresented minorities and persons with disabilities were involved in 225 instances of collaboration. Finally, Space Grant programs were involved in just over 110 partnerships between and among themselves. (The vast number of collaborations cited do not represent the number of projects implemented. Collaborative activities can and do include several partners.)

Total NASA funding for 1994 reached nearly $14 million, with an additional $26.3 million contributed in cash and in-kind matching funds. The bulk of the matching funds, $19.7 million, was contributed nearly equally by lead institutions, academic affiliates, and Federal government organizations.

Education

Cooperative activities and other joint ventures are crucial to the success of state programs, and thus the national program. Industry, government, and academic institutions, along with nonprofit organizations and elementary and secondary schools have formed various alliances to improve science, mathematics, and technology education in their communities. Thus cooperation is an obvious cornerstone of the program. Participants eager to launch new or innovative educational activities would be hard pressed to do so were it not for the time, personnel, facilities and/or funding contributed by other Space Grant partners. Affiliates in the Space Grant program have devised numerous cost-effective methods to broaden the Space Grant program in this time of fiscal restraint. For example, industry members have provided tours of their facilities to teachers, students and faculty, and provided experts to speak at conferences and workshops. Industrial and governmental partners participate in numerous other Space Grant initiatives, including science fairs and career days. The Massachusetts Space Grant Consortium, for example, regularly conducts seminars, using speakers from higher education, industry, and government, to provide students, faculty and the general public with the most current information on aerospace topics.

Industrial and governmental affiliates have contributed to the development of elementary and secondary curriculum, as when the Oklahoma Space Grant Consortium, with the Oklahoma Aeronautics Commission, the Federal Aviation Administration, and Martin Marietta, combined resources to develop a course, using aviation and space science and technology, to enhance teaching skills. In another example, the National Science Foundation taught Oklahoma teachers how to bring electronic meteorological data into Oklahoma classrooms. In Maine, Apple Computers and the Bigelow Laboratory for Ocean Science, a charter member of the Maine Space Grant Consortium, explored how best to use satellite imagery in the classroom. And in Hawaii, the Hawaii Space Grant Consortium, NOAA and NSF distribute rain gauges to interested teachers in the Pacific to conduct climate research. The teachers and their students learn to better understand the spatial structure of tropical rainfall and verify satellite rainfall algorithms.

Both industry and state government offices have contributed research experiences and fellowships to students. Teledyne, Inc., a member of the Missouri Space Grant Consortium, provided a hands-on research activity to top undergraduate engineering students to encourage them to attend graduate school. The focus of their research was the effect of metal catalysts on the thermal stability of ethylene. In Wisconsin, the Wisconsin Geological Survey supported a Space Grant undergraduate research assistant, and in Minnesota, that state's geological survey provided an undergraduate student with a summer research stipend. In Alabama, the NASA Marshall Space Flight Center, Intergraph Corp., and the U.S. Space and Rocket Center, under the auspices of the Alabama Space Grant Consortium, provided laboratory experiences for upper-level undergraduate and graduate students attending universities in Alabama. Two 10-week summer internships, funded by the Indiana Space Grant Consortium and the Boeing Co., were conducted in cooperation with the NSF's Engineering Research Center on Intelligent Manufacturing at Purdue University, lead institution for the Indiana Space Grant Consortium.

Industry and government, in conjunction with Space Grant institutions of higher education, have been instrumental in developing curriculum for the university level. Martin Marietta and the University of Miami, members of the Florida Space Grant Consortium, joined to develop a new interdisciplinary design course. The course served as the basis for one of the six selected Universities Space Research Association's Student Explorer Demonstration Initiative grants. The Indiana Space Grant Consortium and Lockheed Missiles and Space developed a three-semester course on fluid mechanics issues associated with low gravity conditions.

Industry and government partners frequently contribute facilities, equipment, and expertise to Space Grant activities. Students from South Dakota State University, an affiliate of the South Dakota Space Grant Consortium, worked with South Dakota middle school students to design and launch a high-altitude research balloon donated by Raven Industries. Members of the Virginia Space Grant Consortium, along with personnel from FAA and NASA, conducted a competition for undergraduate and graduate students and faculty to develop a multidisciplinary design for a general aviation aircraft. Rust Engineering, a member of the Alabama Space Grant Consortium, assisted students at the University of Alabama in Huntsville in the design of a Get Away Special (GAS Can) experiment. The successful experiment, Corrosion Initiation in Steels, flew aboard the Space Shuttle *Endeavour*, STS 68, in September 1994. Industrial partners have also provided work stations, state-of-the-art computer equipment and software, among many other contributions.

Two recent examples of successful national networking between Space Grant and U.S. government agencies are the Global Observations to Benefit the Environment Program, headed by Vice President Al Gore, and the NASA "Discovery" Program. Space Grant consortia responded rapidly and effectively to a 1994 request for organization of a number of national GLOBE workshops for elementary and secondary teachers. In addition, Space Grant personnel have assisted current and potential Principal Investigators in probing new NASA flight opportunities, such as "Discovery," a recent space exploration initiative. The potential for Space Grant coordination and assistance occurs in numerous existing and planned initiatives related to NASA Strategic Enterprises.[1]

Conclusions

The goal of forming a national network of institutions with interests in aerospace science, engineering and technology has been accomplished; the advantages, each year, become more apparent. In some communities, Space Grant was the force that prompted interaction previously unknown among organizations. The examples noted above are only a sampling of the ways in which Space Grant college and university, industry and government partnerships have benefited U.S. science, mathematics, engineering and technology education. These instances of networking—collaborations, cooperative activities, joint ventures; contributions of facilities, equipment, personnel and the lending of expertise; and the programs that have resulted—have served numerous students of all ages, teachers, faculty and others. What Space Grant institutions and affiliated organizations have shown is that cooperation between and among personnel and institutions provides a prudent and worthwhile opportunity to enrich the U.S. educational experience.

Rocks and Stars: An Interdisciplinary Scientific and Technical Gateway Course

The following, highly personal account of an introductory, interdisciplinary, "gateway" course offers, I believe, an estimate of the potential for the cultivation of scientific and technical literacy within the university/town environment.

Rocks and Stars (R&S) began at Oregon State University in 1981 and was taught once each year until 1988. It was designed strictly to "get" students, in hopes of increasing the Geology Department's staffing; many state universities and other universities receive "formula" funding, based mainly on enrollments. Altruism or any other "quality" reasons were never factors in starting R&S, although I tried to make those cases many times. I pulled out all the stops to promote the course: no prerequisites; science credit; upper division credit; name guest speakers; music; space art; multimedia; humor; informality; lowest possible levels of pretention; dogs and ponies; the works. Pandering? Absolutely. Early on, however, I realized that, to stay out of jail, or at least keep my job, there were some things I couldn't do ("cause"-type things), and there were some things I should do (real grades based on tests). R&S lifted off.

The approach worked beyond anyone's imagination: Enrollment started with 274, the largest geology class ever; in succeeding years, numbers went to 649, then to estimates of more than 1300 each term—classes were restricted to the largest classroom which held 745.

Enrollment was limited to seniors, especially graduating seniors, and juniors, even though the course was designed for freshman.

Imagine, 745 undergraduate students in one room, three time a week. An optional laboratory was added, and more than a third took the lab as well, giving twelve graduate students Research Assistanships each term (popular move, that). All told, R&S accounted for more than 8 FTE via formula funding! Not that we got seven additional staff....

What were some of the innovations? Well, each lecture was taped, some tapes were voice-compressed (total lecture in about 25 minutes), and were available in the language labs, anytime. Students knew that all test questions would come from these tapes. So...miss a week with the tennis team? Sick? Can't get up? No problem— go to the language lab. And, if you wanted to make a good grade, listen up, as much as you might wish. The numbers that did were huge.

Why the popularity? I think the big reasons were lack of pretention and humor. Name guest speakers surely helped. I worked hard at organization, so that each class period was as lively and interesting as possible. Lots of things happened, before and after class (yes, sometimes during class), which could not be termed intellectual, but the 50-minute period was quality.

Success allowed time for reflection. Something good was happening (in addition to increased dollars for the department). It appeared that students were excited about space science and technology, particularly if they were also able to meet people and have a pretty good time. They held 4.6 billion year-old meteorites and Moon rocks in their hands (most geologist haven't). They heard about the most exciting news from top workers in each of the fields—origin of life, possibilities for extraterrestrial life, planetary volcanism, catastrophic asteroid impacts (and the dinosaurs), the fate of the universe.

So? Well, scientific and technical literacy was advanced fairly aggressively. It is hard to prove, but I believe many facts may be remembered a long time; more importantly, I believe perceptions about science and technology were improved, in many cases for students who might not have taken an additional science course. Not bad.

Notes and References

1. Agency Strategic Enterprises are those areas in which the agency will concentrate its efforts. The Strategic Enterprises are: Mission to Planet Earth; Aeronautics; Human Exploration and Development of Space; Scientific Research; and Space Technology.

17

Why Don't Physics Students Understand Physics? Building A Consensus, Fostering Change[1]

Ronald K. Thornton

The benefits of widespread science literacy for non-scientists have been well and eloquently argued. There has been extensive discussion, with some agreement, of the knowledge and skills a person should have to be science literate.[2] Unfortunately less attention has been paid to successful methods of teaching and learning science. There is evidence that traditional methods of teaching science are unable to bring a majority of students, even those intending to become scientists, to understand the physical world. In this discussion I will focus on the teaching and learning of physics.

Are most students in physics courses acquiring a sound conceptual grasp of basic physics principles? For many years physicists teaching basic courses have believed that they are, but those doing research in physics education have been convinced that they are not. Recently, large studies of students' basic conceptual knowledge before and after introductory physics courses have convinced some in the larger community of physics teachers that there is less basic understanding than they had believed. The results of these studies show that students in good universities, students who are able to solve many traditional problems involving algebraic equations or even those requiring the methods of the calculus, fail to agree with physicists when they answer the simplest conceptual questions.

Traditional science instruction in the United States, refined by decades of work, has been shown to be largely ineffective in altering student understandings of the physical world. Even at the university level, students, who take physics courses, whether they be science majors or not, enter and leave the courses with fundamental misunderstandings of the world about them essentially intact: their learning of facts about science remains within the classroom and has no effect on their thinking about the larger physical world. The ineffectiveness of these courses is independent of the apparent skill of the teacher, and student performance does not seem to depend on whether students have taken physics courses in secondary school.

There are a number of barriers to a societally useful science education for non-scientists (and even for potential scientists). Students often perceive science as difficult, boring, and overly concerned with detail. This is due in part to societal stereotypes and in part to the courses actually offered. Science is exciting to scientists because they are engaged in discovery and in creatively building and testing models to explain the world around them. Yet scientists rarely *preach what they practice*. Science courses rarely reflect the practice of science. In most courses, students "do" no science and only hear lectures about already validated theories. Not only do they not have an opportunity to form their own ideas, they rarely get a chance to work in any substantial way at applying the ideas of others to the world around them. The worst courses consist of the presentation of collections of unrelated science facts and vocabulary with no attempt to develop critical thinking or problem solving skills.

Consider traditional instruction in dynamics, force and motion, as an example. Although a Newtonian framework is essential to understanding non-relativistic (and later relativistic) motion, it is common for more than 80% of students to answer most questions from a non-Newtonian point of view after an introductory physics course. Such students may believe, for example, that a net force is required to keep an object in motion at a constant velocity, that

there is a residual force on an object that has been pushed and re-
leased that keeps it moving, and that acceleration must increase as
velocity increases. In contrast, students and physicists who believe
the world behaves in a Newtonian manner (for every day speeds)
use a conceptual understanding based on Newton's laws of motion
and understand that a body moving at constant velocity requires no
net force to keep a body moving and so no residual forces are required. They also understand that a constant acceleration produces
a uniformly increasing velocity. Learning to substitute values into
the equations of motion seldom results in Newtonian conceptual
understanding. Research has shown that standard instruction only
changes the conceptual point of view of 5 to 15% of the students in
the area of dynamics (force and motion). Figure 17.1 shows the

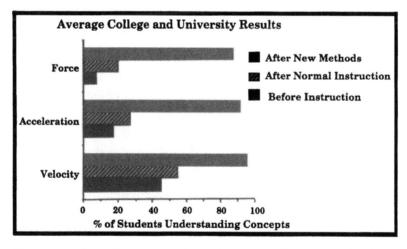

Figure 17.1   Composite data on student understanding of kinematics (labeled Velocity and Acceleration concepts) and dynamics, force and motion as described by
Newton's Laws (labeled Force concepts). Dark bars show student understanding
coming into beginning courses at the university, striped bars are after all traditional
instruction. While the percentage of students who know concepts coming in can vary
with the selectivity of the university, the effect of standard instruction is to change
the minds of only 7 to 15% of students. New methods described in text (lighter solid
bars) result in approximately 90% of students understanding concepts. (Students
evaluated using the *Force and Motion Conceptual Evaluation*)

results of composite research data for thousands of students at universities who took the *Force and Motion Conceptual Evaluation*.[4]

The failure of beginning physics courses to convince students that the Newtonian view of motion makes more sense of the world than their fragmentary childhood views, has broader implications than the fact that the students do not understand force and motion. It calls into question, particularly for students not intending to become scientists, the validity of the scientific process. If science does not make "sense" for students, then there is no good reason to accept conclusions arrived at through the process of science. Of course, some will "accept" science merely on the perceived authority of scientists without any expectation of understanding it. Such a result is clearly not desirable.

If university instruction is changing the conceptual ideas of only 10% of the students, what is happening in high schools? Figure 17.2 shows composite results for high schools similar to university results. Note that students who use the new methods described below in even an average high school, learn concepts much more successfully than university students who experience good traditional instruction at a selective university.

There is more widespread agreement on the effects of traditional instruction than there is on the solutions to the problems of traditional instruction. For some time, the substantial agreements on many major ideas that exist among researchers in physics education have been masked by disagreement over particular ways of defining the nature and current state of student physics learning and over the effectiveness of various pedagogical responses. Such disagreement has too often meant that much work in research and pedagogy goes on as a series of separate efforts, so that projects with the potential to have widespread impact on physics teaching and learning remain isolated. What is needed to change the state of physics education is agreement on a set of underlying principles about the teaching and learning of physics that will support the integration of the work of many different groups into a coherent

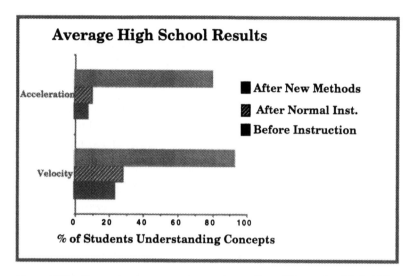

Figure 17.2   Composite data on student understanding of kinematics (labeled Velocity and Acceleration concepts). Dark bars show student understanding coming into physics courses in high schools, striped bars are after all traditional instruction. While the percentage of students who know concepts coming in can vary with the location of the high school, the effect of standard instruction is to change the minds of approximately 10% or less of the students. New methods described in text (lighter solid bars) result in 80 to 90% of students understanding concepts. (Students evaluated using the *Force and Motion Conceptual Evaluation*)

educational response, a response based on careful research, with the potential to have impact on the larger physics and science community.

The beginning of a consensus building process about the results of standard instruction and about pedagogical solutions was made by physicists involved in research in physics education at a meeting entitled "The New Mechanics" at Tufts University in August of 1992. These researchers, who form the New Mechanics Advisory Group, were brought together by Priscilla Laws, David Sokoloff and Ronald Thornton to work toward a number of goals. One purpose of the group's work is to establish general agreement on some methods of teaching of physics that have been shown through research to enhance student learning.

During meetings of the New Mechanics Advisory Group, researchers came to agreement on some generalizations about student learning in physics that were originally drafted by Lillian McDermott. Each generalization is supported by research from different sources using different techniques including, for example, the work of the physics education group at the University of Washington,[3] which elicits detailed accounts of student understanding through interviewing; analysis done by the Center for Science and Mathematics Teaching at Tufts University on responses from thousands of students in introductory physics courses at many different institutions to short-answer questions which are part of the *Force and Motion Conceptual Evaluation*;[4,5] and work done by David Hestenes at Arizona State University in developing benchmark conceptual exams.[6] A list of these points of agreement about students in physics follows.

- Facility in solving standard quantitative problems is not an adequate criterion for functional understanding.
- A coherent conceptual framework is not typically an outcome of traditional instruction. Rote use of formulas is common.
- Certain conceptual difficulties are not overcome by traditional instruction.
- Growth in reasoning ability does not usually result from traditional instruction.
- Connections among concepts, formal representations (algebraic, diagrammatic, graphical), and the real world are often lacking after traditional instruction.
- Teaching by telling is an ineffective mode of instruction for most students.

Each of these generalizations about student learning has strong implications for the changing of physics teaching. It will be difficult for scientists who look at the evidence and who accept these results to find justifications for continuing to teach in a traditional manner. But what is a teacher to do? Even physics education researchers have disagreements about the "best" way to proceed. While most believe that students must be intellectually engaged and actively involved in their learning and that traditional instruction is failing to provide a context in which a majority of students can learn, there is more debate about which methods of teaching and what learning contexts will most help students learn. Can educational technology improve physics learning? Under what conditions does collaborative learning work well? How major a role should experimentation play in student learning?

While the New Mechanics Meeting did articulate agreements about current student learning, the time was too short to agree upon generalizations for suggested methods of physics teaching and because of limited resources, participants were only from the United States. In an effort to see how to start changing our teaching, it may be productive to look at agreements about productive methods of physics teaching reached at an earlier, international meeting of physicists and physics education researchers on a more specific topic: a NATO Advanced Study Workshop entitled *Student Development of Physics Concepts: The Role of Educational Technology* was organized by the author and Robert Tinker and held the University of Pavia in Italy during October of 1989. The participants included researchers from nine different countries and some members of the New Mechanics Advisory Group. The participants were researchers in physics and science education and included major developers of curriculum and pedagogical tools for the teaching of students and teachers. Many of the participants were interested in incorporating recent findings in physics education research and cognitive psychology into new instructional models made possible by the use of interactive technologies.

The NATO workshop was concerned with student conceptual learning and the pedagogical uses of interactive educational technologies in physics teaching and learning. One major focus was on the uses of technologies that allowed students to construct physics concepts successfully from their own experiences of the physical world. Some of the interactive educational technologies demonstrated and discussed at the conference were real-time computer-based data-logging tools (often called microcomputer-based laboratory or MBL tools), the use of robotics for teaching science concepts, interactive video disk/CD-ROM systems, student-directed software pedagogical tools, telecommunication as a means for students to share scientific discoveries, and constructivist intelligent tutor systems.

After examining the evidence, participants were in substantial agreement that students of all ages learn science better by actively participating in the investigation and the interpretation of physical phenomena; that listening to someone talk about scientific facts and results was not an effective means of developing concepts; and that well-designed pedagogical tools (generally computer-based) that allow students to gather, analyze, visualize, model and communicate data can aid students who are actively working to understand physics. In particular, there was evidence from a number of countries (Italy, Germany, UK, USA, USSR) that real-time Microcomputer Based Laboratory tools in appropriate learning environments resulted in successful student learning of physics concepts. It was also agreed that, to best develop their understanding, students need the freedom and ability to pursue interesting scientific investigations, the opportunity to interact with their fellow students, and the means to communicate their findings. (Unfortunately most introductory courses have none of these features.)

The international NATO Workshop resulted in substantial agreement on ways physics teaching could be altered to improve student learning and the conclusions have stood the test of time and research. The New Mechanics meeting at Tufts, in addition to building agreement about student physics learning, began the work of refining curriculum and instructional strategies that will help introductory physics students acquire a conceptual understanding in one specific area of the curriculum—Newton's Laws of Mechanics. The results of this meeting, including a revision of the dynamics sequence described by Arnold Arons in Chapter 2 of his book *A*

Guide to Introductory Physics Teaching, are being incorporated into a new curricular project, *RealTime Physics Mechanics Labs*; and into major revisions of the *Workshop Physics* and *Tools for Scientific Thinking* curricula. Extensive research on learning in these projects will help establish productive instructional techniques and sequences that work for almost all introductory physics students.

Past successes of these projects illustrate the possibility of large learning gains in introductory courses through the use of new methods developed through educational research. For example, the *Tools for Scientific Thinking* guided-inquiry conceptual labs are based on research on what students know, make use of real-time data collection and display (MBL) and encourage students to work collaboratively. By introducing such labs into otherwise conventional introductory courses, it is possible to change student beliefs about the physical world to those held by physicists. After standard instruction, over 1200 students in calculus-based physics courses at five different universities still had a 70% error rate on fundamental acceleration concepts. When, for the first time, two *Tools for Scientific Thinking* active-learning kinematics labs were offered at these universities, more than 75% of students understood these concepts. At universities where there is more experience with the labs, such as the University of Oregon, even in non-calculus introductory courses, 93% of students understand these concepts. At such universities, less than 15% of students held a Newtonian point of view after all traditional instruction in dynamics, while 90% did so after two additional conceptual labs on dynamics. There is good evidence that this conceptual understanding is retained. The *Tools for Scientific Thinking* computer-supported *Interactive Lecture Demonstrations*[4,5] have had similar success in changing the large lecture environment into an interactive environment where students can learn force and motion concepts. Such limited implementation of new methods is not enough, but begins to address the problem of changing instruction in traditional environments. Similar positive results are achieved in the more comprehensive Workshop *Physics* program[7] at Dickinson College, which has replaced lectures with a combination of student-oriented activities using similar active learning techniques and the same educational technology as that described above.

In summary, there is considerable evidence collected by researchers in physics teaching and learning that traditional instructional methods, largely lecture and problem solving, are not effective methods for promoting student learning in physics. There is widespread, but not total, acceptance by researchers of evidence that interactive learning methods, some of which are mentioned above, work well in many different environments. There is enough agreement among careful researchers that the physics community would do well to begin seriously changing traditional teaching methods for all students taking physics and looking in a scientific way at the learning results of these changes.

Notes and References

1. Part of this paper was adapted from Thornton, R.K. "Why Don't Students Learn Physics?," Physics News in 1992, *American Institute of Physics* (1992): 48–50.

2. e.g. Arons, A. "Achieving Wider Scientific Literacy," *Daedalus* 112 (2) (1983): 91–122.

3. e.g. McDermott, L. "How We Teach and How Students Learn—A Mismatch," Proceedings ICPE/IUPAP 4th International Conference, *Teaching Modern Physics: Statistical Physics*, Badajoz, Spain, 1992.

4. e.g. Thornton, R.K. and D. Sokoloff. "Learning Motion Concepts using Real-Time Microcomputerbased Laboratory Tools." *American Journal of Physics* 58 (9), pp. 858–66, Sept. 1990. Thornton, R.K.: "Using Large-Scale Classroom Research to Study Student Conceptual Learning in Mechanics and to Develop New Approaches to Learning." Chapter in book of NATO ASI Series (Berlin-Heidelberg-New York, Springer Verlag) in press. Thornton, R.K. and Sokoloff, D. "Assessing and Improving Student Learning of Newton's Laws Part I: The Force and Motion Conceptual Evaluation and Active Learning Laboratory Curricula for Newton's First and Second Laws," submitted to the *American Journal of Physics*.

5. Sokoloff, D. and R.K. Thornton. "Assessing and Improving Student Learning of Newton's Laws Part II: Microcomputer-Based Interactive Lecture Demonstrations for the First and Second Laws," submitted to the *American Journal of Physics*.

6. e.g. Hestenes, D., M. Wells and G. Swackhamer. "Force Concept Inventory," *The Physics Teacher* 30 (3) (1992): 141–158.

7. Laws, P. "Calculus-Based Physics Without Lectures," *Physics Today*, pp. 24–31, December 1991.

18

Toward A Science-Friendly Society
Loyal Rue

Eric Chaisson has raised the prospect of a "scienceless society," the unfortunate outcome of waning public curiosity about scientific questions and growing public suspicion about the benefits of costly and arcane scientific research. The inevitable question: what good reasons would a society given to such trends have for continuing its support for the scientific enterprise? The inescapable answer: none at all! Indeed, it would be foolish for a society to allocate public resources to expensive projects it fails to understand and believes to be potentially harmful. The scientific community understands this reasoning perfectly well, with the result that there is much hand-wringing in labs these days about the proverbial plug being pulled. All of which brings scientists to ask what can be done to enhance public sophistication about science— the assumption being that to know science is to love it (and thus to leave the plug well enough alone).

On the surface of things there appears to be some irony in the fact that scientists worried over a loss of public support for science have no compelling data to confirm their fears. Public support for science remains positive, as do public attitudes toward scientists.[1] The public, it appears, loves science despite a low level of sophistication about it. If the fears of the scientific community are legitimate (as I believe they are) then it is only because the public's commitment to science is soft and potentially fickle. I will give reasons to support this view momentarily, but for the present I will assume that concerns about the foundations of public support for science are justified.

What, then, to do about it? The danger of this question is that it generates too many tactical responses i.e., specific proposals for reforming science curricula in the public schools. I am certain that tactical issues of curriculum reform will need to be addressed at some point along the way, but I will suggest that these issues might be more profitably engaged in the light of more general concerns. This said, I will focus my remarks on three strategic proposals—*philosophic, programmatic* and *pedagogical*—each of which I believe holds promise for deepening public support for science.

A Philosophic Proposal

I propose that scientists should become very serious about engaging in systematic dialogue with "the other culture" (i.e., humanists), and with representatives of religious communities in particular. The sooner the better. Why is this an important strategic issue? In the experience of many scientists such dialogues have managed only to deepen suspicions about science. So why disturb a sleeping dog? The science/humanities (and especially the science/religion) dialogue is extremely important because it is precisely at those points where scientific knowledge bears upon human self-understanding that public support for science goes soft. Public opinion surveys indicate that a majority (two-thirds) of the American public feels some tension between science and religion. As many as 44% feel a "high level of tension in the form of discomfort or wariness about science, which manifests in a willingness to prohibit certain areas of scientific research".[1] Furthermore, the source of this discomfort has to do with the scientific study of human nature and behavior. It appears that there is little worry about science (thus high support for it) as long as the focus of inquiry is limited to the likes of rocks and stars. But when human nature and behavior become the focus of scientific inquiry, then fears begin to rise. This concern becomes especially urgent when one considers the issues brought forth by current brain research, one of the hottest domains in science these days.

Why is it that humanists and religious believers become starchy when science is brought to bear on questions of human nature and behavior? It is because humans have a need to affirm the intrinsic value of their humanity, and this need is well served by some theory or another about *human uniqueness*. Squeeze a humanist or a religious believer hard enough and you will get the reply that scientific approaches to human reality tend to compromise the affirmation of human uniqueness. I stress the point that uniqueness *itself* appears to be at stake here, more than any particular theory about it. Atheist humanists are just as excitable on this point as religious fundamentalists. But of course the tension becomes most real when a particular theory of human uniqueness is brought to point by some line of scientific inquiry. When this happens it becomes clear that humanist/religious theories of human uniqueness are normally based on wisdom traditions that predate the emergence of science, and these traditions (often mythological) usually stand in sharp contrast to the concepts and categories of science, thus the discomfort.

It may be contended that nothing in science precludes the claim for human uniqueness. Quite the contrary—scientists are themselves just as inclined to affirm the uniqueness of humanity as anyone else, though perhaps on different grounds. But this is precisely why scientists should engage the other culture in dialogue. It is very likely the case that science can provide more and better legitimations for human uniqueness than ancient wisdom traditions can. And to the extent that these legitimations are made intelligible, science will be received as a welcome resource rather than being resisted as a threat.

One should not gloss over the difficulties involved in the dialogue being proposed here. Scientific understandings of human nature and behavior are not compatible with *every* humanist or religious perspective, any more than these perspectives can be reconciled among themselves. But for starters it may be enough for scientists to insist vigorously that science is not even slightly hostile to one of the deepest principles underlying these perspectives. Given this opener, and a commitment to sustaining the dialogue over the long haul, the source of public discomfort about science will begin to deconstruct.

A Programmatic Proposal

Secondly, I propose that those concerned with the effectiveness of science education consider alternative ways to think about what counts as scientific sophistication. The National Science Board presently employs the following set of true/false questions as a measure of general scientific literacy,

1) The oxygen we breathe comes from plants.
2) The center of the earth is very hot.
3) The continents are moving slowly on the face of the earth.
4) Light travels faster than sound.
5) All radioactivity is not manmade.
6) The earth goes around the Sun once each year.
7) Electrons are smaller than protons.
8) The earliest humans lived at the same time as the dinosaurs.
9) The universe started with a big explosion.
10) Lasers are not composed of sound waves.

I am not about to offer a critique of these indicators of scientific literacy. They seem quite adequate for their purpose, which is to generate comparative data on who knows what about selected terms and concepts used in science. But what these indicators fail to tell us is whether an individual has actually internalized the modern scientific world view. In the following I will record my own subjective standards for judging scientific sophistication. I suspect the following indicators would be found to correlate rather well with a firm positive regard for the scientific enterprise.

I propose two indicators for judging scientific sophistication, the first having to do with a general perspective and the second with habits of mind. The first indicator would measure the extent to which an individual understands the narrative of cosmic evolution as it is expressed by the physical, biological and social sciences. This narrative account presents us with a universe that began some 15 billion years ago with a big bang, a universe that developed through the formation of galaxies and the generation of stars, wherein the elements of chemical composition were forged. The story of cosmic evolution further records the assembly of our solar system, the evolution of the earth's formations, and the origin and diversification of life forms. This story locates human existence in the context of biological evolution and ecological succession. It tells us that humans are one species with a common origin, and that we have diversified into a broad range of cultural and psychological orientations by virtue of historical accidents.

The second indicator of scientific sophistication is a measure of the extent to which one has acquired certain habits of mind that correspond roughly to the methods of science. These will include:

1) a drive to render one's experience intelligible
2) a disposition to seek the truth without regard for the outcome
3) a reticence to form opinions without a broad base of evidence
4) a willingness to change one's opinions in response to new evidence
5) a preference for the simplest adequate explanations
6) an aversion to "taking someone else's word for it."

I grant that these indicators may resist quantification, and may therefore be useless for the purpose of generating comparative data. Nevertheless, they appear to come close to what we mean by a modern scientific outlook. My guess is that very few persons who have internalized the perspective of cosmic evolution and who exude scientific habits of mind could also manage to harbor suspicions about the scientific enterprise.

The virtue of these intellectual characteristics as programmatic principles is that they establish the narrative framework and the mental attitudes that are conducive to lifelong learning in science. The habits of mind dispose individuals to be active and critical seekers of new knowledge, while the narrative of cosmic evolution provides a coherent structure for assimilating and organizing virtually everything they learn. If an educational system were to succeed in producing these two general outcomes then there would be little worry over the prospect of a scienceless society.

A Pedagogical Proposal

The standard critique of public school systems is that the teachers haven't learned anything worth teaching. This is especially true in the case of elementary teachers whose college level preparation is saturated with courses on methods and management, with only incidental course work in subjects with teachable content. The consequence is that instruction amounts to little more than a guided tour through textbooks which the teachers themselves too often have difficulty understanding. Under these conditions the subjects with greater objectivity of content (i.e., math and science) are often the most difficult and the least interesting, because they lend themselves to rote methods. I well remember the anxiety my own children experienced as they prepared for science tests. They were normally reduced to memorizing lists of vocabulary items which had relevance for them only with respect to the impending exam. When the exams came along they performed reasonably well, and the annual skills assessments showed that the children had achieved respectable levels of mastery in science. And yet none of my children ever enjoyed science for the very reason that they could never see the *point* of it.

For years I was left mystified by these children who would spend whole afternoons laboring in search of rocks and fossils, or tinkering with inventions in the garage, but who never took as much as a passing interest in science. I now believe they were turned off to science because it was too far removed from their experience. (There is a single exception to this general observation: *the bone unit*. Each of my three children has vivid memories of their grade school science unit on the bones of the human body, the one unit that managed to play to their experience.)

If we genuinely aspire to a science-friendly society then we might consider the proposal that elementary schools discontinue the teaching of science altogether. No more science before high school. Period. Instead, let the schools concentrate on providing students with meaningful experiences with natural phenomena. Focus on nature and forget about science. The most constructive elementary school policy I can think of would be to get rid of science textbooks and to hire an enthusiastic naturalist, someone who hasn't a clue about lesson plans and assessment strategies, but who can excite and nurture a child's powers of observation. Real estate agents often quip that property values depend on three things: location, location and location. In a similar way, the value of primary science education depends on observation, observation and observation.

Access to natural phenomena in the company of a competent naturalist who has a reasonably good chance of answering childish questions is all the science instruction one could hope for in the early grades. Let the children discover in high school that there is such a thing as science, a systematic body of knowledge about such stuff as they have come to appreciate on their own terms. Give the public early childhood programs like this and they will be far more likely to develop and sustain an interest in science.

A science-friendly society, is one in which science functions as a reliable public resource for understanding the natural world and our place within it. If we have worries about our society becoming unfriendly toward science then there are several countervailing steps that might be taken to render science less threatening, more meaningful and more interesting. I have proposed that science might become less threatening to humanistic perspectives if scientists were to help in articulating a more plausible theory of human uniqueness. Secondly, I have suggested that attempts to enhance public sophistication in science should have less to do with principles and facts than with world views and habits of mind. And finally I have proposed that elementary science education should begin where science begins—with careful observations of natural phenomena.

Notes and References

1. Miller, D. Jon, "Science and Religion: The Impact of Religious Tension About Science on Public Attitudes Toward Science and Technology." A paper presented to the American Association for the Advancement of Science, Chicago, February 14, 1987.

Other titles in World Futures General Evolution Studies